Maria-Lluisa Gea-Valor, Isabel García-Izquierdo
& Maria-José Esteve (eds)

Linguistic and Translation Studies in Scientific Communication

PETER LANG
Bern · Berlin · Bruxelles · Frankfurt am Main · New York · Oxford · Wien

Bibliographic information published by Die Deutsche Nationalbibliothek
Die Deutsche Nationalbibliothek lists this publication in the Deutsche Nationalbibliografie; detailed bibliographic data is available on the Internet at ‹http://dnb.d-nb.de›.

British Library and Library of Congress Cataloguing-in-Publication Data:
A catalogue record for this book is available from *The British Library*, Great Britain.

Library of Congress Cataloging-in-Publication Data

Linguistic and translation studies in scientic communication /
Maria-Lluisa Gea-Valor, Isabel García-Izquierdo & Maria-José Esteve (eds).
p. cm. – (Linguistic insights ; ISSN 1424-8689 ; v. 86)
Includes bibliographical references.
ISBN 978-3-0343-0069-8 (alk. paper)
1. Science–Language. 2. Communication in science. I. Gea Valor, M. Lluïsa.
II. García Izquierdo, Isabel. III. Esteve, María José.
Q226.L55 2010
501'.4–dc22

2010019812

ISSN 1424-8689
ISBN 978-3-0343-0069-8

Hochfeldstrasse 32, CH-3012 Bern, Switzerland
info@peterlang.com, www.peterlang.com, www.peterlang.net

Printed in Switzerland

Linguistic and Translation Studies
in Scientific Communication

Linguistic Insights

Studies in Language and Communication

Edited by Maurizio Gotti,
University of Bergamo

Volume 86

PETER LANG
Bern · Berlin · Bruxelles · Frankfurt am Main · New York · Oxford · Wien

Contents

Maria-Lluïsa Gea-Valor / Isabel García-Izquierdo /
Maria-José Esteve

Introduction

Over the past few years, the field of *English for Specific Purposes (ESP)* has become fairly consolidated. Within this broad term, further areas of interest have been traditionally distinguished: *English for Academic Purposes (EAP), English for Business and Trade (EBT), English for Science and Technology (EST), English for Legal Purposes (ELP)* and *English for Vocational Purposes (EVP)*. This volume specifically deals with the *EST* branch which, as its name indicates, is concerned with the specific uses and functions of language in scientific and technological disciplines. Although language in science has traditionally been valued for prioritising objective and propositional content, interpersonal and pragmatic dimensions have lately been explored with many interesting results. In addition to the linguistic perspective, relevant contributions have also been made within the field of Translation Studies which facilitate the understanding of specialised communications.

Taking this into account, the volume *Linguistic and Translation Studies in Scientific Communication* aims to further contribute to the development of the *EST* field by offering insightful perspectives from linguistics and translation. More specifically, the papers in the book cover topics such as popularisation and transmission of scientific knowledge via the technologies of information and communication; terminology and corpus-based studies in scientific discourse; genres and discourse in scientific and technical communication; history and evolution of scientific language; and translation of scientific texts.

The book is divided into two sections, which mirror the two perspectives above-mentioned: the first part is entitled 'Construction and Communication of Scientific Knowledge'; and the second one, 'Translation and Scientific Knowledge'.

Section I opens with Martin Hewings's chapter, where he explores the impact of the popular press on the image of science and scientists as well as the differences in the way this image is projected to the general public. He analyses two science news articles published in the British tabloid *The Sun*, tracing their journalistic history back to the original research papers. By focusing on the language of science reporting in the press, Hewings sheds light on the complex interaction between specialist and popularised communication in today's society.

Chapter 2, by FRANÇOISE SALAGER-MEYER, offers a diachronic exploration of scientific discourse by studying the rhetorical evolution of criticism, which is essential to the very construction of scientific knowledge. The paper focuses on the genre of the academic book review (BR) and analyses a range of features closely related to the critical task in a corpus of medicine BRs in three distinct periods: the closing years of the 19th century, the mid 20th century and the beginning of the 21st century. Salager-Meyer emphasises that only by making reference to the history of science will we be capable of understanding the formal and rhetorical features of today's science.

In the next paper, LUZ GIL-SALOM and CARMEN SOLER-MONREAL explore the appraisal resources most frequently used in the discussion sections of research articles (RAs) in four engineering disciplines: Computing, Telecommunications, Nanotechnology and Robotics. The authors focus on evaluative, certainty and epistemic expressions, which most conspicuously reveal the writer's attitude and engagement towards propositions in the text and, more importantly, towards the reader.

Chapter 4, by CONCEPCIÓN ORNA-MONTESINOS, deals with the semantic relations of hyponymy (the 'kind of' relation) in the specialised discourse of architecture, construction and civil engineering. More specifically, the set of cognitive synonyms, hyponyms and meronyms provided by the lexical database of the WordNet project for the noun *building* is contrasted with the frequency analysis of its hyponyms in the *Construction Textbooks Corpus* (CTC). Results show how semantic relations condition the choices available to the members of this discourse community to lexicalise their disciplinary world.

NURIA EDO, in chapter 5, offers a selection of entries from the prospective specialised bilingual (English-Spanish, Spanish-English) dictionary of the ceramics industry she is currently working on. Within the framework of contrastive lexicography, the terminological units of the ceramics industry field are shown in a useful and realistic manner, emphasising the communicative and associative nature of such units.

Chapter 6, by Inés Lareo, presents the main guidelines of a long-term project carried out by the research group *Multidimensional Studies in English* at the University of A Coruña. This project, the *Coruña Corpus of Early Scientific Writing,* comprises several subsections of scientific areas in the Modern English period, especially Astronomy. The CETA *(Corpus of English Texts on Astronomy)* subcorpus is the object of study of this researcher and is also focused on in chapters 7 and 8.

In chapter 7, Gonzalo Camina examines the processes of word formation in twenty samples of the Philosophy section of the *Coruña Corpus of Early Scientific Writing*, paying special attention to affixation. Camina is particularly interested in how science coined new nouns in the Early Modern English period.

This first section of the book finishes with Estefanía Sánchez Barreiro's paper, in which she studies the occurrences of the so-called 'extenders' in a collection of scientific texts from the 18th century. The main interest of her study is to provide an overview regarding the use of this type of constructions in the formal context of scientific discourse, a fact which poses a challenge since extenders are more frequently found in informal and/or oral communication.

In Section II, Tomás Conde's paper presents a study on serial translation evaluation, focusing on the differences in the behaviour of four groups of evaluators – potential addressees, professional translators, translation teachers and translation students – while appraising specialised and non-specialised texts. Conde concludes that there are two aspects that influence the differences observed: the evaluators' self-confidence about the subject matter, and the nature and disparate complexity of the technical and the non-specialised prose.

Chapter 10, by Pilar Ordóñez, provides a revision of the characteristics of specialised communication, placing particular emphasis on legal translation. Drawing on the concept of textual genre, the author describes each stage of the translating process as performed by a specialised legal translator, using the multilingual GENNT *Corpus of Specialised Genres.*

In chapter 11, Francisca Suau-Jiménez draws attention to the importance of metadiscourse in the translation of scientific texts. She carries out a pilot study which analyses hedges and phatic elements, drawing on a corpus of RAs and popular science articles in English and Spanish. Suau-Jiménez concludes that the main differences between English and Spanish metadiscourse are cross-generic rather than intra-generic. Nevertheless, the need to undertake deeper and wider research in order to explore frequency of these features in both languages and genres is highlighted.

Olga Torres-Hostench, José Ramón Biau, Pilar Cid, Adrià Martin, Bartolomé Mesa-Lao, Mariana Orozco and Pilar Sánchez-Gijón describe the design and development of the *TRACE (Traducción Asistida, Calidad y Evaluación*) research project, which is currently investigating the impact of CAT (Computer-Assisted Translation) tools on target text production. Drawing on the hypothesis that the final texts obtained using CAT tools differ from those in which such tools have not been used, the research group carried out an experiment which involved professional translators working in a simulated real-world environment. The experiment established specific variables applied to the study of technical texts.

In chapter 13, Juan José Martínez-Sierra draws on the dramatic increase of scientific and technical texts in the media, especially documentaries on television. Documentaries are defined as an audiovisual genre (in contrast with written and oral genres) in which scientific and technical contents find their place, and which prioritise specific translation modalities, namely dubbing, voice-over and subtitling. Martínez-Sierra points out that audiovisual translation makes it possible for documentaries originally produced in another language to cross the language barrier and to be screened in Spain. Finally, some implications for translators are discussed.

The closing chapter of the volume, by María Rosario Bautista Zambrana, explores the topic of ontologies for scientific and technical translation in the medical field of diabetes. For this purpose, she resorts to the Termontography approach, and to the methodology for ontology creation, the so-called 'Methontology'.

We hope the rich and interesting variety of approaches offered in *Linguistic and Translation Studies in Scientific Communication* will provide useful insights, and ultimately contribute to a better understanding of scientific communication today.

Section I. Construction and Communication of Scientific Knowledge

Martin Hewings

'Boffins Create "Supermouse"': The Role of the Popular Press in Creating the Public Image of Scientists and their Work

1. Introduction: the public image of science and scientists

This study explores the image of scientists and their work that is projected in science journalism and, in particular, considers differences in the way this image is projected in publications aimed primarily at non-specialists. An analysis is conducted of two science news articles that appeared in the British tabloid *The Sun*, tracing their journalistic history back to the source academic papers. These two examples are taken from an ongoing, more extensive comparison of science articles in the British tabloid press and texts reporting the same stories in other publications.

The public image of science and scientists is an area of considerable research interest, dating back to a pioneering study conducted by Mead/Metraux in the USA in the mid 1950s (Mead/Metraux 1957). In this study, the authors found a complex image of scientists among high school students. On the one hand, informants reported an 'official' version of how scientists should be perceived:

> The scientist is seen as being essential to our national lives and to the world; he is a great, brilliant, dedicated human being, with powers far beyond those of ordinary men, whose patient researches without regard to money or fame lead to medical cures, provide for technical progress, and protect us from attack. We need him, and we should be grateful to him (1957: 387).

On the other hand, elicitation of a more personal response from informants produced a rather more negative composite image, which Mead/Metraux summarised as follows:

> The scientist is a man who wears a white coat and works in a laboratory. He is elderly or middle aged and wears glasses. [...] He may be bald. He may wear a beard, may be unshaven and unkempt. He may be stooped and tired. [...] He is surrounded by equipment. [...] He spends his days doing experiments. [...] He writes neatly in black notebooks (1957: 386-387).

Since this time, a body of research has grown up investigating images of scientists among both children and adults, mainly accessed through their drawings (Finson 2002). Chambers (1983), for example, using the Draw-a-Scientist Test (DAST) with elementary school children, found that the features normally associated with this stereotypical image were: a male, a lab coat (usually white), glasses, facial hair, scientific instruments and laboratory equipment, books and filing cabinets, technology or the products of science (such as rockets), and captions such as formulae. More recently, a survey of public attitudes towards science in Britain uncovered a perception that:

> scientists and engineers are "not quite like us," that perhaps they operate according to a different moral code, being driven by the desire to discover and create while not necessarily pausing to think about the consequences (Office of Science and Technology and the Wellcome Trust 2000).

Overall, then, it might be concluded that: "the stereotypical image of a scientist is alive and well" (National Science Foundation 2002). Scientists are seen not to be like ordinary people, and part of this image is unattractive to the public. Although many scientific developments are welcomed, some of their work is viewed as having little value, and scientists may be unaware of the consequences of their research. Potentially, therefore, they are threatening and dangerous to society.

There are two main reasons why the public image of science and scientists should be of concern and, therefore, an area of academic enquiry. One is that public image can influence the career choices of young people. If we assume that societies need scientists in order to develop, then their ability to attract young people into post-school scientific education is crucial. Mead/Metraux's (1957) conclusions were, in general, rather discouraging in this respect. Essentially, they observed a conflict between (what were then) contemporary values in society and the image of the scientist. An unwillingness among young people to associate themselves with the negative image of scientists is obvious, but Mead/Metraux went on to say that even "the positive image of very hard, only occasionally rewarding, very responsible work is also one which, while it is respected, has very little attraction for young Americans today" (1957: 387). The concern in some societies about the lack of attractiveness of science as a career continues until the present day. In recent years in Britain, for example, there have been closures and mergers of university science departments, at least partly as a consequence of difficulties in recruiting students. Frean (2007), for example, contends that:

> A shortage of science graduates threatens the future of British industry at a time of record demand for scientists to combat problems such as climate change and to take advantage of trends in the global economy. [...] Since 1984 the number of people studying physics A level has slumped by 57 per cent, and the take-up of chemistry has dropped by 28 per cent [and] the proportion [of university students] graduating in physics and chemistry fell by 25 per cent between 1994 and 2006.

A second main reason is that the image can influence the effectiveness of communication by scientists. Scientists are responsible, whether through news media, television documentaries, public lectures, or other forums, for communicating information to the public about scientific issues. There is a broad consensus that it is important for citizens and their leaders to be scientifically literate (Miller 2004) so that they can both appreciate the role of science in society and be part of an informed debate about its applications, benefits and limitations (Office of Science and Technology and the Wellcome Trust 2000, 2001). The issue goes to the heart of a successful democracy: some level of understanding of stem cell research, genetically modified foods, climate change, and so on, are necessary if we are to assess and influence the policies of governments concerning the applications of scientific knowledge. Clearly, the image of scientists has an impact on the effectiveness of their communication with the public. Put simply, "the more appealing the image, the more likely that people will listen to what is being said" (National Science Foundation 2002) and, it can be added, that if people listen, the more likely they are to be better informed and able to weigh evidence.

2. Influences on the public image of science and scientists

It is impossible, of course, to establish precisely how the public image of science and scientists is created. For many people, their experience of science in school will be an important influence, while other educational experiences such as public lectures or demonstrations of science, or museum visits may play their part. Beyond educational contexts, the cultural models to which people are exposed can contribute to the mental schema they have of the scientist (Gardner 1980). These cultural models derive from multiple sources, such as television, films, and literature. On television, the scientist is often portrayed as amoral (rather than immoral), insensitive and obsessive (Schibeci 1986). Hollywood films have depicted scientists as socially inadequate and lacking practical intelligence while, sometimes at the same time, being a threat to humanity (Terzian/Grunzke 2007). Certainly, in comparison with other professions, the entertainment industry appears to represent scientists in a bad light:

> professions such as medicine, law, and journalism [are presented] as exciting and glamorous, whereas scientists and engineers are almost always portrayed as unattractive, reclusive, socially inept white men or foreigners working in dull, unglamorous careers (National Science Foundation 2002).

Also of significance is the press, which is the focus of the investigation presented here. Some would, in fact, argue that the press is the most significant source of science information for most people. Conrad (1999: 285), for example, contends that:

> Science journalism is an increasingly important wellspring of public understanding of science. [...] For most people, the reality of science is what they learn from the press. In a sense, science journalists become gatekeepers for the infusion of scientific information into the public sphere.

This may overstate its significance. Respondents in a more recent survey (National Science Foundation 2002) in the USA identified television (44%) as their main source of information about science and technology, followed by newspapers (16%), magazines (16%), and the

Internet (9%). It is likely, of course, that the Internet has become a more important source of information since then, but given that an increasing number of newspapers are appearing online, the significance of newspapers (both paper and online versions) may well have increased.

Even if the press is only one of a number of major influences, then the analysis of newspaper texts reporting science is important because the ways in which journalists select and present their stories influence the way science is presented to their readers and, consequently, the image these readers have of scientists and their work.

3. Science reports in the press

The literature on the communication of science is vast (see Weigold 2001, for a review), and attention here will only be on the language of science reporting in the press. Much of the work in this area falls under the broad heading of the study of 'popularisations', characterised by Giannoni (2008: 212) as "the dissemination of scientific findings outside the communities that produce and to a certain extent 'own' such knowledge". According to Paul (2004) the dominant view sees scientific popularisations as often flawed translations of scientific research for a lay audience, but this view has been challenged by those (e.g. Whitley 1985; Myers 2003; Fahnestock 2004) who see specialist and popularised communication interacting with one another in a complex process of mutual knowledge construction (Myers 2003: 270). Cassidy (2005: 115-116) observes, for example, that:

> popular science may play an active role in the constitution of knowledge, where new and controversial science appears in the mass media before it has done so in academic fora such as journals or conferences, or where scientific controversies appear in the public domain.

It has been argued, therefore (e.g. Office of Science and Technology and the Wellcome Trust 2000), that it is more useful to think of an "engagement model" of science communication, involving a two-way dialogue between specialists and non-specialists.

The aim of much of the work from both perspectives has been to identify linguistic evidence of similarities and differences in the vocabulary, grammar, syntax and organisation of research articles and popularisations. Myers (1990), for example, proposes a distinction between a "narrative of science" in professional research articles and a "narrative of nature" in popularisations. In a narrative of science, articles "follow the argument of the scientist, arrange time into a parallel series of simultaneous events all supporting their claim, and emphasize in their syntax and vocabulary the conceptual structure of the discipline" (1990: 142). In popularisations, on the other hand, articles "present a sequential narrative of nature in which the plant or animal, not the scientific activity, is the subject, the narrative is chronological, and the syntax and vocabulary emphasize the externality of nature to scientific practices" (1990: 142).

Macdonald (2005) compares nouns – particularly as sentence subjects – in science research articles and health science reports in journalistic texts. She finds that journalists choose a different repertoire of nouns, reflecting a focus on human interest angles and human actors. For example, she notes that while in research articles attributions typically come at the beginning of the sentence in the form "X reported Y", in journalistic accounts the order is reversed with the attribution coming within or at the end of a quotation. This word order "allows more dramatic material to be put up front [...] and it moves the obligatory attribution out of the way of the drama" (2005: 286).

Most such studies of popularisations have analysed publications such as *Science* and *Nature* which publish across a wide range of scientific fields and include non-specialists in their readership (e.g. Fahnestock 2004), "intermediate publications" (Green 1985) such as *New Scientist* and *Scientific American* aimed primarily at the interested lay reader[1] (e.g. Myers 1990), or science features in general-interest magazines and 'quality' newspapers (e.g. Macdonald 2005). Fewer studies have investigated science reporting in the tabloid press, even though in terms of circulation the tabloid press has the potential to

1 *New Scientist* says that it sees its readership as "scientists and non scientists as a way of keeping track of scientific and technological progress". Source: <http://www.newscientist.com/info.ns?id=in16> last accessed 3rd October 2008.

influence a much larger number of people. For example, two of the publications explored in the research below are the British tabloid *The Sun*, published six days a week, and the weekly *New Scientist*. *The Sun* has sales of around 3 million daily with an estimated readership of close to 8 million,[2] while *New Scientist* sells something around 170,000 weekly with about 800,000 readers worldwide.[3] Readership figures for *The Sun* are in fact likely to be considerably higher as it is also available free online. *New Scientist* has an online version, but this can be accessed only by subscribers.

The research reported below compares images of scientists and their work projected in intermediate publications (*New Scientist*), the quality press (*The Independent*), and the tabloid press (*The Sun*) with particular focus on *The Sun*.

4. The study

Story 1: The Large Hadron Collider

The first story concerns the Large Hadron Collider (LHC) – the 'atom-smasher' built at the CERN particle physics centre in Geneva in Switzerland – which was switched on in September 2008. The articles examined were published in late 2007 and early 2008, and reported speculation on what would happen when the LHC was switched on. (At the time, it was believed that this would take place in May 2008.) The original research paper, on which the later article in *The Sun* was based, was published online on 25th October 2007 in the *International Journal of Geometric Methods in Modern Physics* with the title "Time Machine

2 Average daily sales between 29th December 2008 to 25th January 2009 were 3,146,006. Average daily readership between January and June 2008 was 8,011,000. Source: <http://www.nmauk.co.uk/nma/do/live/factsAndFigures?newspaperID=17> last accessed 17th February 2009.

3 Average weekly sales were 172,820 and average global readership 838,670 between 1st January and 30th June 2008. Source: <http://www.newscientist.com/data/html/ns/mediacenter/uk/intro_mag.jsp> last accessed 17th February 2009.

at the LHC",[4] and written by two Russian mathematicians, Igor Volovich and Irina Aref'eva. Essentially, these writers speculated on whether the LHC could tear a hole in space and time, forming a wormhole, potentially allowing particles from future time to pass through it. It is perhaps worth noting the highly speculative nature of their work. An extract from the conclusion comprises three *if*-clauses:

(1) We argued that if one can trust the classical geometrical estimate of the cross-section for the black hole production, if there exists an exotic matter similar to the dark energy, and if the scale of quantum gravity is of the order of [a] few TeVs then one can expect the production of time machine/wormholes in the proton-proton collisions at the LHC of the same order as the cross section for the black hole production.

and the final sentence calls for further studies:

(2) Further studies of the experimental signatures of the wormhole production are required since there are transitions between black holes and wormholes.

A related *New Scientist* article was published on 7th February 2008 with the title "2008: Does time travel start here?"[5] The opening paragraph of the article is given here:

(3) AS YOU may have heard, this will be the year. The Large Hadron Collider – the most powerful atom-smasher ever built – will be switched on, and particle physics will hit pay-dirt. Yet if a pair of Russian mathematicians are right, any advances in this area could be overshadowed by a truly extraordinary event. According to Irina Aref'eva and Igor Volovich, the LHC might just turn out to be the world's first time machine.

A related article in *The Sun* was published on 9th February 2008 with the title "Visits from crack to the future".[6] The first four paragraphs are given here:

4 <http://arxiv.org/abs/0710.2696v2> last accessed 17th February 2009.
5 <http://www.newscientist.com/article/mg19726421.700-2008-does-time-travel-start-here.html> last accessed 17th February 2009.
6 <http://www.thesun.co.uk/sol/homepage/news/article772800.ece> last accessed 17th February 2009.

(4) We could soon be visited by TIME TRAVELLERS – thanks to a £2billion machine able to punch a HOLE in the universe.
The atom-smasher may open up a time tunnel for people from the future to travel back to now.
The particle accelerator, called the Large Hadron Collider, has been designed to explore the origins of the universe by investigating the tiniest components of atoms.
It is due to be switched on in May.

Perhaps the most obvious difference between the *New Scientist* and *The Sun* articles is the balance between informing and entertaining, with entertainment clearly weighted more heavily in *The Sun*. The headline, typical of *Sun* style, is a play on words, echoing the film title *Back to the Future*. The journalist's position is given as "Sun Spaceman". The article is illustrated with a photograph showing the arrival of a time traveller, and it gives five tips on how to spot one: he's won the lottery five weeks in a row, he spends his summer holidays in 2076, and so on.

In comparing the texts from *New Scientist* and *The Sun* I briefly consider three aspects: *accuracy of information, attribution,* and *how scientists are represented.*

Accuracy of information

The *New Scientist* article is an attempt to represent very complex ideas in a form that will be accessible to non-specialists. This necessitates simplification and less detail than is found in the original. I assume, though, that any inaccuracies are either an almost inevitable consequence of this simplification process, or a consequence of the lack of understanding on the part of the writer, rather than a deliberate attempt to misrepresent the original. In *The Sun*, however, we find something apparently rather different. Referring back to the opening paragraphs of *The Sun* article, given above, the implication is that in May, visits from time travellers become a possibility. This claim, it is said, comes from one of the writers of the original paper:

(5) Dr Volovich, of the Steklov Mathematical Institute in Moscow, reckons future technology may be advanced enough to send someone back through the time tunnel. *(The Sun)*

The *New Scientist*, however, explicitly rules this out in:

(6) they [Volovich and Araf'eva] are not actually suggesting that visitors from the future are imminent. *(New Scientist)*

and there is no suggestion of this in the original paper.

Clearly, it is only a tiny minority of *Sun* readers who might actually have believed that time travellers from the future would start arriving in Geneva in May. And the purpose of *The Sun* article is not to warn readers of this possibility: it is, rather, to derive entertainment from scientific news. Most readers will find it mildly amusing and dismiss it as "Nonsense" or "A bit of fun". But, of course, scientists are portrayed as having come up with this "nonsense": so implicitly they are criticised for doing something of no value and therefore wasting time and, presumably, taxpayers' money.

Attribution

Both *New Scientist* and *The Sun* report the authors in indirect reports and using quotations. The indirect reports in *New Scientist* appear to be paraphrases, including clarifications, of the original paper. For example:

(7) Causality is one of [sic] fundamental physical principles. We suggest in this note that there is a possibility to test [sic] causality in experiments at CERN's Large Hadron Collider (LHC). *(Original paper)*

(8) What they are saying is that since causality – the idea that effect must follow cause – is one of the most fundamental principles of physics, the notion that it might be tested at the LHC is worth pushing as far as possible. *(New Scientist)*

It is not clear where the quotations in the *New Scientist* come from. They may have been elicited from the scientists themselves, or they may have been part of a press release or some other publicity document. They include:

(9) "We realised that closed timelike curves and wormholes could also be a result of collision of particles," Aref'eva says. *(New Scientist)*

In *The Sun* article, there are three sections where words are attributed directly or indirectly to Volovich and Aref'eva. In at least two of these, however, the words reported appear actually to be those of the *New Scientist* journalist. Consider the following section of the *New Scientist* article in which there is no attribution:[7]

(10) Yet who knows? Perhaps future civilisations might work out how to stabilise and grow a wormhole, then manipulate the two mouths in order to *create a time tunnel. If a combination of fast-moving particles and phantom energy does create a wormhole* in Geneva this year, such *an advanced civilisation could find it in their history books, pinpoint the moment, and take advantage of their technology to pay us a visit. (New Scientist)*

This is picked up in the following extract from *The Sun* article:

(11) Dr Volovich, of the Steklov Mathematical Institute in Moscow, reckons future technology may be advanced enough to send someone back through the *time tunnel.* [...] He told the magazine New Scientist: "*If a combination of fast-moving particles and phantom energy does create a wormhole, an advanced civilisation could find it in their history books, pinpoint the moment – and take advantage of their technology to pay us a visit*". *(New Scientist)*

There is no mention of a "time tunnel" in the original article, but it appears to have been introduced by the *New Scientist* as a way of giving readers a visual image by which they might better understand an abstract idea. *The Sun* makes this more concrete, suggesting that people could be sent back through the tunnel, for some readers conjuring up images from the television series *The Time Tunnel.* In the second part of the extract, Dr Volovich is quoted directly. However, these are not actually Volovich's words, but rather a close paraphrase of something written by the author of the *New Scientist* article.

At best, we might excuse this as lazy journalism. At worst, this process of second- or third-hand reporting misrepresents not only the original research but also the legitimate attempt of *New Scientist* to make this research accessible to a wider readership.

7 In this and in following extracts, language of particular interest is highlighted in italics.

Representation of scientists

There is one other section in *The Sun* article in which scientists are reported. Much of:

(12) And *some boffins think it may be possible for people of our time to travel years ahead* – like TV Timelord Doctor Who. *(The Sun)*

appears to be based on the following section of the *New Scientist* article:

(13) The possibilities this raises are being taken seriously by *some physicists*. "This is an interesting paper," says J. Richard Gott of Princeton University in New Jersey, *who suggested* as long ago as 1991 that *accelerating particles could be a route to time travel. (New Scientist)*

As can be seen, J. Richard Gott, an eminent professor of astrophysical sciences, is labelled in *The Sun* a "boffin". In addition, a photograph is included of one of the authors of the original paper, above the title "Big Bang boffin – Dr Volovich". The photograph includes two of the stereotypical features of the scientist: (thick lens) glasses, and baldness.

The term 'boffin' is used mainly in British colloquial English for "an intellectual, an academic, a clever person; an expert in a particular field; *esp.* such a person perceived as lacking practical or social skills" (Oxford English Dictionary). To explore the use of the term further in *The Sun* and compare its use in *New Scientist*, two sub-corpora from the Bank of English[8] are analysed. *The Sun* corpus totals 44.76 million words and comprises issues of *The Sun* and its Sunday sister paper, *The News of the World*, from 1998 to1999. The *New Scientist* corpus totals 7.89 million words and comprises issues of *New Scientist* from 1991 to 1998. Table 1 gives the number of occurrences of *boffin(s)* in these two corpora. Figures in brackets give the number per million words for ease of comparison.

---

8 I would like to thank Collins Dictionaries and the University of Birmingham for permission to use and quote from the Bank of English.

	The Sun	*New Scientist*
boffin	112 (2.5)	13 (1.6)
boffins	236 (5.3)	27 (3.4)
Total	348 (7.8)	40 (5.1)

Table 1. Number of occurrences of *boffin(s)* in the corpora.

To examine the connotations of the term in these two corpora, it is useful to look at qualitative adjectives used to describe *boffins* and, for comparative purposes, *scientists*. In *The Sun* corpus boffins are *clever* and *creative*, but also *unlucky, bungling, dozy, eccentric, nerdy,* and *cardigan-wearing*. Scientists are *well-informed, world-renowned, self-taught, sexy, young*, but also *corrupt, evil, loony, maniac, oddball,* and *stuffy*.

Given some of these characteristics associated with 'boffins' it may seem surprising that the word is used only a little less frequently in the *New Scientist* corpus than in *The Sun* corpus. In fact, the majority of these are references to the public's perception of scientists as 'boffins', usually with a tone of complaint or despair, as in:

(14) physics makes a huge contribution to wealth creation in Britain, and is not simply an abstruse hobby pursued at vast public expense by hair-brained *boffins* in ivory towers.

And in one case, there is specific mention of the image of scientists in the press:

(15) The tabloid press is the worst offender, still perpetuating the myth that scientists are "*boffins*" who make "amazing discoveries".

In other examples, 'boffin(s)' is used in a humorous way, with reference, for example, to "alien boffins". This is also reflected in the way that the word 'boffin' is used creatively in *New Scientist*, with occurrences of *boffindom, boffinesque, boffinitis, boffinocracy* and *boffinry*, none of which are found in *The Sun* corpus.

Overall, use of the label 'boffin' and the language associated with it in *The Sun* contributes to the image of the scientist as clever but unworldly. In *New Scientist*, on the other hand, the term is introduced mainly in order to complain about or satirise its use.

Story 2: Boffins create supermouse

The second article from *The Sun* illustrating the portrayal of scientists is entitled “Boffins create ‘Supermouse’”. [9] The research reported in the article was published originally in the *Journal of Biological Chemistry* in a paper entitled “Over Expression of the Cytosolic Form of Phosphoenolpyruvate Carboxykinase (GTP) in Skeletal Muscle Repatterns Energy Metabolism in the Mouse”[10] in November 2007. On 1st November there was a press release from the researchers’ home university – Case Western Reserve University in the US – headed “Case Western Reserve researchers breed a mighty mouse”. [11] On 2nd November an article appeared in *The Independent* (a ‘quality’ British newspaper) entitled “The mouse that shook the world”.[12] On 3rd November *The Sun* picked up the story in its article. (The same story was published in many other newspapers, too.) *The Independent* article included extensive quotations, mainly from the lead researcher, Professor Richard Hanson. Roughly half of *The Sun* article is quotation from Professor Hanson, two other sentences were taken from the press release, and the rest was (apparently) based on *The Independent* article.

Here, extracts from the earlier sources (the press release and the article in *The Independent*) will be compared with *The Sun*’s reporting of the story. The press release included the following section:

(16) *[they] have the capability of running five to six kilometers at a speed of 20 meters per minute on a treadmill for up to six hours before stopping*. These genetically engineered mice also *eat 60 percent more than controls, but remain fitter, trimmer* and *live and breed longer than wild mice in a control group. Some female PEPCK-*

---

9 <http://www.thesun.co.uk/sol/homepage/news/article423596.ece> last accessed 17th February 2009.

10 <http://www.jbc.org/cgi/content/abstract/M706127200v1> last accessed 17th February 2009.

11 <http://blog.case.edu/case-news/2007/11/02/mightymouse> last accessed 17th February 2009.

12 <http://www.independent.co.uk/news/science/the-mouse-that-shook-the-world-744870.html> last accessed 17th February 2009.

> *Cmus mice have had offspring at 2.5 years of age, an amazing feat considering most mice do not reproduce after they are one year old.* Animal behavior studies later demonstrated that the PEPCK-Cmus mice are seven times more active in their home cages than controls; in addition, the mice were also markedly *more aggressive*. *(Press release)*

A number of elements of this section of the press release were picked up and reworked in the article in *The Independent*:

(17) *The mouse can run up to six kilometres (3.7 miles) at a speed of 20 metres per minute for five hours or more without stopping.* Scientists said that this was equivalent of a man cycling at speed up an Alpine mountain without a break. *Although it eats up to 60 per cent more food than an ordinary mouse, the modified mouse does not put on weight. It also lives longer and enjoys an active sex life well into old age – being capable of breeding at three times the normal maximum age.* "On the downside, they eat twice as much as control mice, but they are half the weight, and are *very aggressive*. Why this is the case, we are not really sure". *(The Independent)*

and again in *The Sun*:

(18) MEET Supermouse – the sex-mad mutant with the power of a superhero. The genetically modified rodent runs for almost six hours without a break, eats like a horse and still has enough energy for a few randy romps afterwards. [...] They live longer than other mice and have an insatiable sexual appetite which lasts well into old age. But it also flies into uncontrollable rages which have baffled their creators at the Case Western University in Cleveland, Ohio. *(The Sun)*

The exaggeration is clear. The fact that they breed for a longer period makes them "sex-mad" with "an insatiable sexual appetite which lasts well into old age", and the observation that they are "more/very aggressive" becomes their tendency to fly into "uncontrollable rages".

In addition to exaggeration, *The Sun* attributes human characteristics – albeit excesses – to the mice. This contributes to its main slant on the story: that the technique might be applied to humans. Although neither the original article nor the press release made any reference to this possibility, the article in *The Independent* raises the question, and quotes one of the researchers:

(19) Could the science that created this supermouse be applied to humans? [...] Professor Hanson said: "We humans have exactly the same gene. But this is not something that you'd do to a human. It's completely wrong. We do not think that this mouse model is an appropriate model for human gene therapy. It is currently not possible to introduce genes into the skeletal muscles of humans and it would not be ethical to even try". *(The Independent)*

The Sun, however, suggests that it is a more imminent possibility:

(20) This has led some critics to point to the dark side of this scientific breakthrough. They fear the same technology could be applied to *HUMAN BEINGS* with disastrous effects. Rogue researchers could potentially alter the same gene in the human race, leading to super-strong men and women who stay slim for life but are plagued by a raging lust. *(The Sun*; capitals and highlighting were in the original article)

Although it is not stated who these "critics" might be, later in *The Sun* article, a quotation is included from the head of the organisation *GeneWatch*, [13] and it is this person, Helen Wallace, who is presumably the "critic":

(21) Last night *GeneWatch* chief Helen Wallace warned that the science must *NOT* be turned on humans. She said: "If the aim of this research is to create a new breed of super humans this is disturbing. Any attempt would involve completely unjustifiable and dangerous experiments on mothers and their babies". *(The Sun;* capitals and highlighting were in the original article)

It seems unlikely that anyone, including the scientists involved in the research, would disagree with Helen Wallace. However, by labelling the quotation as a warning, *The Sun* is suggesting that there is something to be warned against: in other words, it implies that Hanson and his colleagues might be working towards this aim.

---

13 According to its website (<http://www.genewatch.org> last accessed 23rd May 2008), *GeneWatch* is "a not-for-profit group that monitors developments in genetic technologies from a public interest, environmental protection and animal welfare perspective". Helen Wallace, the Executive Director of *GeneWatch UK* "has a degree in physics from Bristol University and a PhD in applied mathematics from Exeter University [and] has worked as an environmental scientist in academia and industry and as Senior Scientist at Greenpeace UK".

The Sun article seems to be feeding on, and in turn contributing to, an image of the scientist as someone who is not to be trusted with the knowledge that they have – someone who is beyond society's control. This image is projected in other, perhaps more subtle ways, in the article. The press release begins with the following extract, in which I have highlighted verbs which indicate processes in the research:

(22) CLEVELAND--Case Western Reserve University researchers *have bred* a line of "mighty mice" [...] Parvin Hakimi, the article's lead author and a researcher in the Hanson lab, *developed* this new line of PEKCK-C mice over the past five years as part of on-going research [...] The transgenic mice, which now number nearly 500, *were derived* from six founder lines [...] *(Press release)*

In contrast, in *The Sun* article, emphasis is on the *creation* of the mice:

(22) Five hundred of the mini-marvels *have been created* after boffins in America modified their energy-burning DNA [...] their *creators* at the Case Western Reserve University in Cleveland, Ohio. The first mega-mouse *was created* four years ago by injecting a highly active form of a gene into a mouse embryo. Prof Hanson and his team have spent the past four years *creating* more and analysing their activities. *(The Sun)*

When it is used in connection with bringing life forms into existence, "create" seems to be a loaded term: God creates life, but people create monsters. This is reinforced in the article by the mice being labelled "mutants":

(23) The *mutants* have been revealed for the first time in the Journal of Biological Chemistry. *(The Sun)*

The word "mutant" is not used in the original article, the press release or *The Independent*. In fact, technically, this seems to be a correct use of the term in one of its senses: an organism that is different from others of its type because of a permanent change in its genes. However, most non-specialist readers are more likely to associate it with mutants found in comic books, films and science fiction writing in which they are mutated life forms with superhuman abilities. Often these seem to be the product of exposure to radiation after nuclear war or scientific experiments that have gone wrong.

This implicit suggestion that the scientists were doing something that they should not manifests itself elsewhere in *The Sun* article, too. According to the press release, the unusual characteristics of the mice came about as a result of the "over-expression" of the gene:

(24) According to Hanson, the key to this remarkable alteration in energy metabolism is the *over-expression* of the gene for the enzyme phospheonolpyruvate carboxykinase (PEPCK-C). *(Press release)*

In *The Independent*, it was the result of a "modification":

(25) the animals came about as a result of a standard genetic *modification* to a single metabolism gene shared with humans. *(The Independent)*

while in *The Sun* it was the consequence of the scientists' "tampering":

(26) The key to the success of the experiment was *tampering with* an energy-burning gene – humans have the same one – to make the mouse rapidly burn off calories while never getting muscle cramps. *(The Sun)*

'Tampering' is usually associated with someone trying to change something that they should not, perhaps because they do not know enough about the thing, or because they are trying to damage it.

Overall, then, the language choices made in *The Sun* article combine to project an image of the rogue scientist, beyond the control of the restraining powers of society and involved in experimentation to produce something unnatural and frightening. Their ultimate motive is not to develop mutant animals, but to use the knowledge they gain to produce super-humans.

5. Discussion

5.1. Summary of observations

The exploratory analysis I have presented here has briefly compared the image of scientists and their work in the British tabloid press, as represented in *The Sun*, with that in the 'intermediate' publication *New Scientist*, and the 'quality' newspaper *The Independent*.

Differences are clear. In reporting these scientific investigations, *The Sun* appears to exaggerate and distort the researchers' goals and achievements. Scientists are projected as being at best frivolous, and at worst having unworthy motives. The image of the scientist as 'not like us' is reinforced; they are labelled 'boffins' who are focused on their research to the exclusion of all else. The comparable texts in *New Scientist* and *The Independent* are more measured, apparently paying more attention to accurately reporting the original research and scientific processes.

Given the traditions of the tabloid press in Britain, it is perhaps not surprising that *The Sun* gives entertainment a greater weighting than informing in its reporting of scientific investigation. Like most newspapers, and quite reasonably, it looks for a human interest angle to its reports of scientific work: what impact will results of this work have on readers? However, in the texts I have looked at, *The Sun* goes beyond human interest to sensationalism, encouraging an emotional reaction to science news in a way that is not found in *New Scientist* or *The Independent.* The underlying message seems to be to warn its readers about the consequences of scientific developments, creating the impression that scientists are not to be trusted with the knowledge that they produce.

There is, of course, no simple direct relationship between how the print media represents scientists and their work and public perceptions of them. However, intuitively, it seems likely that if a publication creates a consistent network of associations attached to scientists and their work, and these are repeated frequently, then these associations will form part of how the readers of that publication will perceive them.

5.2. Problem and response

Much of the research on the public perception of science, including the work reported here, assumes that the public image of science to which the press and other media contribute is a 'problem'. One recent manifestation of this concern was the launch (on 28th January 2009)[14] by the British government of a campaign to change public attitudes to science following the publication of a survey showing that most people feel that science is remote, elitist and irrelevant to their lives. In setting up the campaign, better communication and education are seen as ways of helping to change perceptions.

An additional step that might be taken is to find ways of encouraging the public to become more critical readers of science journalism. More might be done in school and other contexts to encourage an informed, sceptical approach to the information in such texts. Below I report on two initiatives which provide models for what might be done. One is a column, and related website, in *The Guardian* newspaper by Ben Goldacre, and the other comes from the website of the British National Health Service (NHS).

Ben Goldacre's column 'Bad Science' in *The Guardian* (see <http://www.badscience.net> and Goldacre 2008) sets out to uncover inaccuracy and misrepresentation of science, particularly in the media. To illustrate, in a recent article (Goldacre 2009), he examines how scientific research on the effects of drinking coffee were reported in the British popular press. The message conveyed by *The Daily Express* was fairly typical:

> Too much coffee can make you hallucinate and sense dead people, say sleep experts. [...] People who consume lots of caffeine in a typical day are three times more likely to experience the freaky hallucinogenic effect when they try to sleep than those who stick to just the odd cuppa (Blacklock 2009).

Goldacre points out that this figure ("three times more likely") does not appear in the original research paper, but:

> seems to be an ad hoc calculation done afterwards by the researchers, and put into the press release, so you cannot tell how they did it, or whether they controlled appropriately for problems in the data.

---

14 <http://news.bbc.co.uk/1/hi/sci/tech/7854121.stm>

He goes on to say that:

> putting a finding in the press release but not into the paper is a subversion of the peer review process. People will read this coverage, they will be scared, and they will change their behaviour. But the researchers' key reported claim, with massive popular impact, was never peer reviewed, and crucially the technical details behind it are not in the public domain.

In addition to warning readers that the report in *The Daily Express* was probably an inaccurate representation of the original research, he suggests that in producing press releases, scientists (and their organisations and sponsors who might have a part to play in writing them) need to consider carefully how the information contained within them might be used by the popular press.

The 'Behind the headlines' section of the British NHS (<http://www.nhs.uk/News/Pages/NewsIndex.aspx>) is aimed at the public, health professionals and journalists. The articles it contains "explain the facts behind the headlines and give a better understanding of the science that makes the news". In most of these articles, a recent health news story in the press is analysed and compared with the research paper on which it was based. The articles not only point out discrepancies between press reports and the original research papers, but also give a commentary on and evaluation of the original research on which the reports are based. To illustrate, there follow two extracts from recent articles, the first on the relationship between sleep loss and weight, and the second on the relationship between eating eggs and cholesterol levels:

> The Daily Express may have given the impression that all sleep disorders can be solved by weight loss. This is not the case, as not all sleep disorders are due to obstructive sleep apnoea. There are many other reasons for disturbed sleep that have nothing to do with weight.
>
> (<http://www.nhs.uk/news/2009/02February/Pages/Sleepproblemsandweight.aspx>)

> As this research does not describe its methods, it is not possible to carry out a full appraisal of its quality. A systematic review would be needed to better establish egg safety. However, it is highly likely that the original research which advised limiting egg consumption was faulty.
>
> (<http://www.nhs.uk/news/2009/02February/Pages/Eggsafety.aspx>)

6. Concluding comments

The work I have reported here is very much an exploratory study of just two sets of texts, and clearly much more needs to be done in considering the impact of the popular press on the public image of science. A further limitation of the work is its focus on only the British context. Consistent international comparison of the public perception of science in different countries can be extremely difficult, although it has been reported (Office of Science and Technology and the Wellcome Trust 2000: 30) that compared with New Zealand and the USA in 1997 and Japan in 1995, Britain in 2000 is less positive about science than these countries. However, British people are less hostile towards scientists compared with Japan in 1995. Of particular current interest would be attitudes to science in rapidly industrialising countries such as Brazil, China and India, where society has a pressing need to encourage more young people into the science and engineering professions. The influence of the press in creating a positive or negative image of scientists and their work in such countries would be a fruitful area of investigation.

References

Blacklock, Mark 2009. Danger from Just 7 Cups of Coffee a Day. *The Daily Express*. 14 January 2009. <http://www.express.co.uk/posts/view/79820/Danger-from-just-7-cups-of-of-coffee-a-day>

Cassidy, Angela 2005. Popular Evolutionary Psychology in the UK: An Unusual Case of Science in the Media? *Public Understanding of Science*. 14, 115–141.

Chambers, David 1983. Stereotypic Images of the Scientist: The Draw-a-Scientist Test. *Science Education*. 67, 255-265.

Conrad, Peter 1999. Uses of Expertise: Sources, Quotes, and Voice in the Reporting of Genetics in the News. *Public Understanding of Science*. 8, 285-302.

Fahnestock, Jeanne 2004. Preserving the Figure: Consistency in the Presentation of Scientific Arguments. *Written Communication*. 21, 6-31.

Finson, Kevin 2002. Drawing a Scientist: What We Do and Do Not Know after Fifty Years of Drawings. *School Science and Mathematics*. 102, 335-345.

Frean, Alexandra 2007. 'Golden Carrots' to Lure Students to science. *Times Online*, August 13, 2007. <http://www.timesonline.co.uk/tol/news/uk/education/article2246470.ece>.

Gardner, Howard 1980. *Artful Scribbles*. New York: Bavi Books.

Giannoni, Davide 2008. Popularizing Features in English Journal Editorials. *English for Specific Purposes*. 27, 212-232.

Goldacre, Ben 2008. *Bad Science*. London: Fourth Estate.

Goldacre, Ben 2009. Bad Science: The Things You Can Perk Up with a Cup of Coffee. *The Guardian,* 17 January 2009, 16.

Green, Jeremy 1985. Media Sensationalisation and Science: The Case of the Criminal Chromosome. In Shinn, Terry / Whitley, Richard (eds) *Expository Science: Forms and Functions of Popularization*. Dordrecht: D. Reidel, 139-161.

Macdonald, Susan Peck 2005. The Language of Journalism in Treatments of Hormone Replacement News. *Written Communication*. 22, 275-297.

Mead, Margaret / Metraux, Rhoda 1957. Image of the Scientist among High-school Students. *Science*. 126, 384-390.

Miller, Jon 2004. Public Understanding of, and Attitudes toward, Scientific Research: What We Know and What We Need to Know. *Public Understanding of Science*. 13, 273–294.

Myers, Greg 1990. *Writing Biology: Texts in the Social Construction of Scientific Knowledge*. Madison, WI: University of Wisconsin Press.

Myers, Greg 2003. Discourse Studies of Scientific Popularization: Questioning the Boundaries. *Discourse Studies*. 5, 265-279.

National Science Foundation 2002. *Science and Engineering Indicators – 2002* (Chapter 7: Science and Technology: Public Attitudes and Public Understanding). National Science Foundation, Division of Science Resources Statistics: Arlington, VA. <http://www.nsf.gov/statistics/seind02/c7/c7h.htm>

Office of Science and Technology and the Wellcome Trust 2000. *Science and the Public: A Review of Science Communication and Public Attitudes toward Science in Britain.* Office of Science and Technology / The Wellcome Trust.

Office of Science and Technology and the Wellcome Trust 2001. Science and the Public: A Review of Science Communication and Public Attitudes toward Science in Britain. *Public Understanding of Science.* 10, 315-330.

Paul, Danette 2004. Spreading Chaos: The Role of Popularizations in the Diffusion of Scientific Ideas. *Written Communication.* 21, 32-68.

Schibeci, Renato 1986. Images of Science and Scientists and Science Education. *Science Education.* 70, 139-149.

Terzian, Sevan / Grunzke, Andrew 2007. Scrambled Eggheads: Ambivalent Representations of Scientists in Six Hollywood Film Comedies from 1961 to 1965. *Public Understanding of Science.* 16, 407-419.

Weigold, Michael 2001. Communicating Science: A Review of the Literature. *Science Communication.* 23, 164-193.

Whitley, Richard 1985. Knowledge Producers and Knowledge Acquirers: Popularisation as a Relation between Scientific Fields and their Publics. In Shinn, Terry / Whitley, Richard (eds) *Expository Science: Forms and Functions of Popularization.* Dordrecht: D. Reidel, 3-28.

Françoise Salager-Meyer

Academic Book Reviews and the Construction of Scientific Knowledge (1890-2005)

1. Yesterday and today: the importance of diachrony

Everything is *diachronic*: nothing is static. Everything changes, moves, evolves, whether it is inorganic or organic matter. Everything is past and project. What does not move or change simply ceases to be.

In linguistics, diachronic studies have been making an important come back in the past 20 years or so. Anglo-American linguists and sociologists have given various names to the historical study of language and discourse, such as socio-historical linguistics (Romaine 1982), historical sociolinguistics (Milroy 1991), diachronic text linguistics (Fries 1983), linguistic history (Granda 1980), new philology (Fleischman 1990) and, more recently, historical discourse analysis (Brinton 2001) or historical pragmatics (Jucker 1995). Whichever the label adopted, however, all of them underscore that there is a close relationship between text and context – whether it is social, economic, political or cultural, and that knowledge evolves and is socially and historically constructed.

More specifically, in the field of languages for specific purposes (LSP), Bazerman (1988) claims that it is necessary to understand why certain forms of expression arise, why they evolve and why they disappear. As a matter of fact, he further stresses that it is only through the analysis of language as a historical phenomenon that one can understand the rhetoric of today's scientific discourse. Hyland (2000) echoes this opinion by asserting that diachronic analyses of scientific discourse are an important way of looking at the constitutive relationship between writers and their texts in the sense that they show that language and communities are mutually entailed and constituted.

Using many of the procedures of rhetorical and register analysis, diachronic research, such as that conducted by Bazerman (1988), Atkinson (1996), Valle (1997) or Salager-Meyer (2004), has been able to reveal important changes in several academic genres – although mainly in the sacrosanct research article – and to underline how the writers' rhetorical strategies have changed as a response to the needs, conceptions and practices of the scientific community.

All in all, LSP diachronic studies have shown that if the scientific research article is to be understood and analyzed as a genre, it should not be inferred from it that it is a stable genre with unchanging and immutable laws and conventions. On the contrary, a brief review of its development as a genre helps reveal how its form and style have changed over time and how its textual evolution has addressed the communicative purposes of scientists as researchers and writers as well as the discourse functions of journal articles as required by the scientific-academic community (Kuo 1999).

2. Science and criticism

Through the analysis of the evolution of certain quantitative and qualitative features of a relatively understudied genre – the academic book review (BR) – I would like to illustrate how instructive the diachronic exploration of scientific discourse can be. I shall make particular emphasis upon the rhetorical evolution of a socio-pragmatic phenomenon that is "an integral feature of the book review genre" (Hyland 2000: 61) and that represents "the common thread that holds the genre together" (Giannoni 2002: 356), a phenomenon without which, it is important to underline, there would be no science at all. I am referring here to *criticism*, an element that is essential to the dynamic process of scientific knowledge construction.

The concept of criticism is not new at all. On the contrary, back in 1852, the Irish physician, Sr. Dominic Corrigan (1802–1880) was already underlining the importance of academic controversy in the following terms:

> Whether my observations and opinions be disproved or supported, I shall be equally satisfied. Truth is the prize aimed for: and, in the contest, there is at least this consolation, that all competitors may share equally the good attained (Cited in O'Brien 1980: 1356).

Closer to us, structuralism and deconstructivism both stress the fact that scientific knowledge is relative, thus positing that criticism – hence refutability (Karl Popper, William James) – are essential components of the scientific movement, i.e., that both are the motor of scientific paradigm evolution.

As the French philosopher, Jacques Bouveresse (1984: 45) claims:

> Tant que la critique réelle ne sera pas considérée absolument normale et indispensable, et le désaccord des adversaires de bonne foi quelque chose de plus intéressant et productif que le consentement des partisans, notre philosophie sera malade.[1]

Indeed, in order to maintain alive a scientific thought from whichever discipline, it is necessary to submit it to discussion and revision because whoever says *science* or *research* also says refutation, argumentation and contra-argumentation, hypothesis formulation, doubt, disagreement and, above all criticism. We can thus posit that science is the construction of arguments and counter-arguments that we try to fit together in a mental puzzle, and that the linguistic formulation of these arguments depends on historically and culturally determined contexts.

3. The construction of scientific knowledge

I shall illustrate my arguments with examples taken from BRs published in English-medium medical journals in three distinct periods: the closing years of the 19th century, the mid 20th century and the beginning of the 21st century.

---

1 "Until criticism be considered as something absolutely normal and necessary, and the disagreement of good faith opponents as more interesting and productive than our partisans' consensus, our philosophy will be ill". [My translation].

3.1. End-of-19th century (1890-1900)

3.1.1. General features

a) Book review by-lines: general information (Table 1, rows 1 and 2)

As can be expected, the book title, the author's name, the number of pages and of illustrations, the name of the publishing house, the year of publication and, from 1899 on, the book price were found in the book review by-lines.

	End of 19th century	Mid 20th century	Beginning 21st century
1. Details about the book	Title, author(s), number of pages, illustrations and plates, publishing house and year of publication. From 1899 on: price	Title, author(s), price, number of pages and illustrations, publishing house + address, year of publication	Title, author(s), number of pages, publishing house, ISBN, price, illustrations, book size, publishing house phone number and web page.
2. Types of books and number of authors	authored-books:[a] 72% 1 author: 70% 2 authors: 2% 3 or more: 0% Edited books:[b] 16% 1 editor: 10% 2 editors: 6% 3 or more: 0% NA: 12%	authored-books: 64% 1 author: 44% 2 authors: 14% 3 or more: 6% Edited books: 30% 1 editor: 18% 2 editors: 6% 3 or more: 6% Published by organisations: 6%	authored books: 42% 1 author: 24% 2 authors: 10% 3 or more: 8% Edited books: 58% 1 editor: 22% 2 editors: 20% 3 or more: 16%

a An 'authored book' is a book that has been written by 1, 2, 3 or more authors, the chapters of which are written by these same authors.

b An 'edited book' is a book edited by 1, 2, 3 or more editors, the chapters of which are written by different contributors.

3. Author's identification	Very detailed: title, academic degrees, member of scientific societies, present and past positions, etc.	Name and surname; sometimes academic degrees held	Initial of first name + surname
4. Book reviewer's identification	Anonymous (62%) Initials only (37%) First name initial + surname (1%)	First name initial + surname (92%) + sometimes academic status (8%)	First name initial, surname, workplace, medical specialty and address (100%) Rarely: 2 reviewers.
5. Languages	German (70%), English (30%)	English (96%), German (4%)	English (100%)
6. Book review length	Total N words: 58,800 Mean: 1,176 Range: 58 - 4,854 -Very short (a-crítical) -Very long with verbatim quotes and digressions	Total N words: 26,400 Mean: 528 Range: 139 - 854 -Very short (a-crítical) -Other longer ones	Total N words: 16,250 Mean: 324 Range: 254 - 538 'Book note': 322 More homogeneity
7. Criticism tone of voice	Direct, harsh, very personal	Polemical, aggressive, sarcastic, humorous, face threatening	Casual, neutral, monotonous, conciliatory, matter-of-fact
8. Targets of criticism	Medical praxis incorrect nomenclature/ classification Visual elements	Conceptual: omission of information, trivial/inexact Contextual: book price; audience Typographical errors and errors of interpretation (tables, graphs)	Coherence: lack of coherence Between-chapter redundancy; poor editing; lack or unrecency of bibliographical references Errors: web page and statistics

9. The judge	*We, us*: an expert A prolific, direct and polemical 'judge' *Science-in-the-making*	*I, My opinion*: an ordinary reader A less verbose but still polemical, aggressive and sarcastic judge	Absent or *I, In my opinion, I disagree* An ordinary, humble, conciliatory and consensus-seeking reader *Science once made*
10. The accused	A subject/person: the author book clearly identified by his surname	Both a person and an object: the book author rarely identified by his surname (*The author*) or the book itself.	Object: the book or a book chapter. Impersonalisation of the accused
11. Global criticism	Frequent (in final position)	Sometimes (final position)	Always ends up with some sort of positive remark

Table 1. Recapitulatory Table.

As can be seen from Graphs 1 and 2, 72% of the books reviewed were authored books[2] whereas only 16% were edited works. Graph 1 also shows that 70% of the authored books were written by a *single* author.

2 I shall refer to books written by one or more authors as *authored books* and to books written by one or more editors as *edited books/works*.

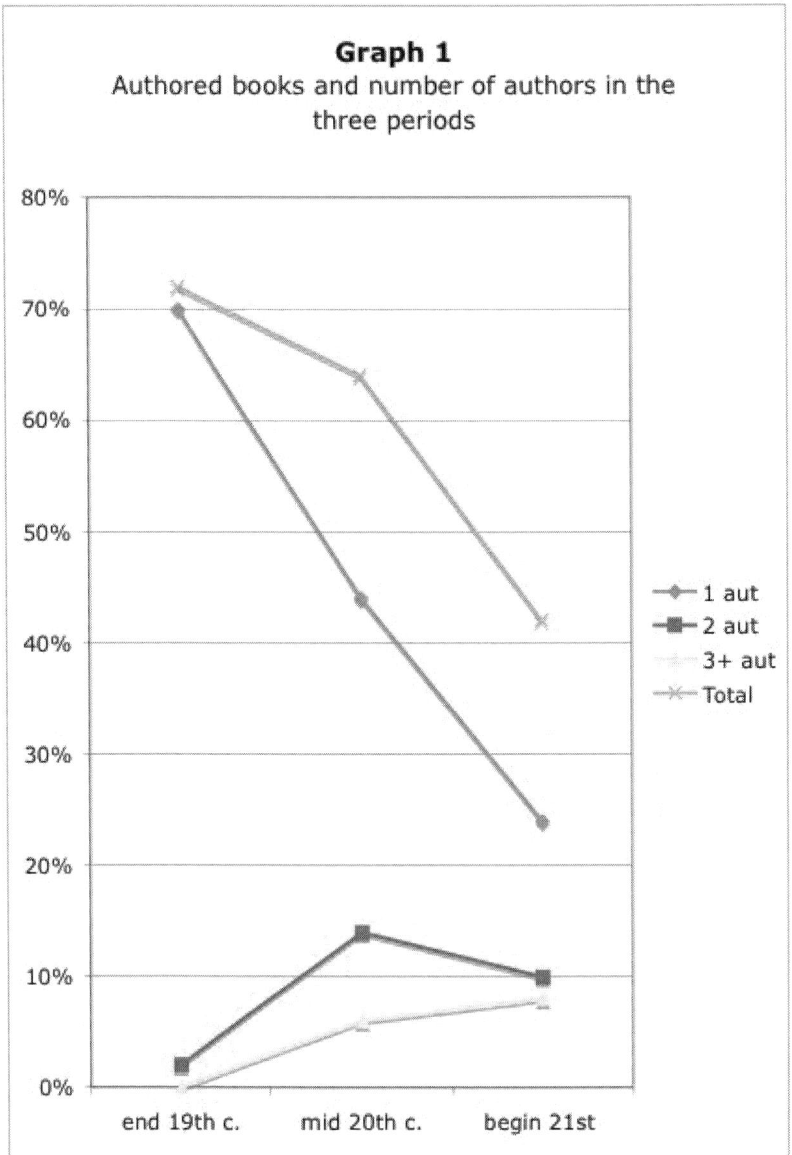

Graph 1. Authored books and number of authors in the three periods.

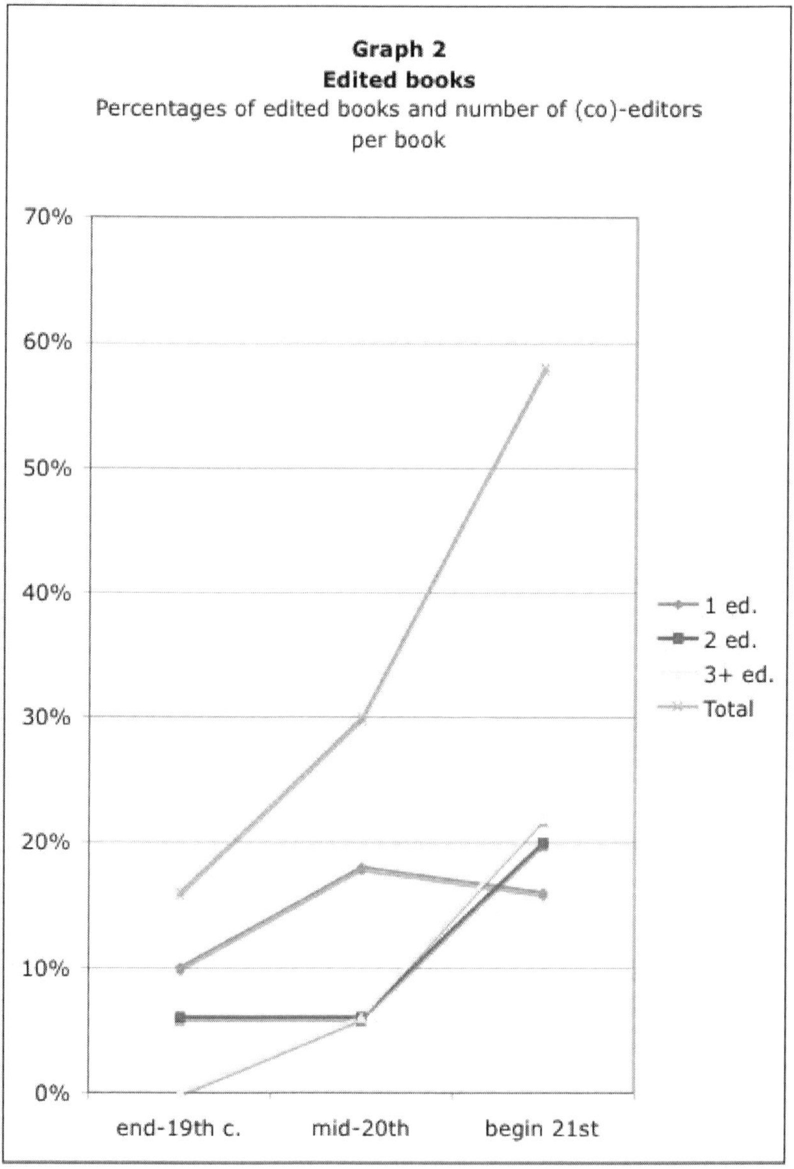

Graph 2. Percentages of edited books and number of (co-)editors per book.

b) The book author and the book reviewer (Table 1, rows 3 and 4)

The book author was indicated along with very detailed biographical information (e.g., academic titles, membership to scientific societies, present and past academic positions, etc.) to the point that, when the list was too long, it ended up with 'etc'. Here is a typical example of the time (year 1893):

> Obstetric Nursing. By Fleetwood Churchill, M.D., formerly President of the Royal College of Physicians in Ireland. F.R.C.S. Ed., Obstetric Physician and Gynecologist, Mater Misericordiae Hospital; ex Examiner, Conjoint Board Royal College of Surgeons and Apothecaries' Hall; Consulting Physician, Hospital for Children, etc. etc.etc.

By contrast, the book reviewer (Table 1, row 4) was most of the time anonymous (62%) or simply mentioned with the initial of his first name and that of his surname (37%).

c) Languages of publication (Table 1, row 5)

70% of the books reviewed were originally written in English vs. 30% in German (see Graph 3), the majority of these being read in the original language by the book reviewer. Only a few of the books written in German were read in their translated (English) version.

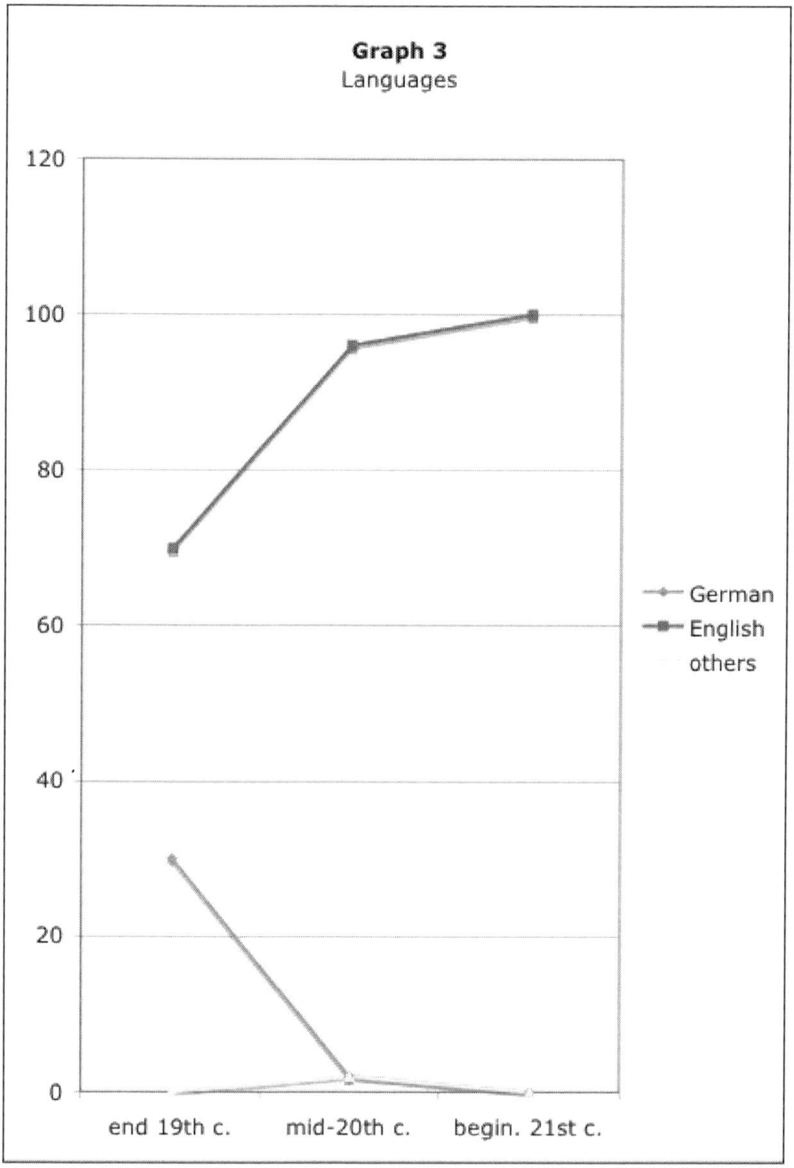

Graph 3. Languages.

d) Book review length (Table 1, row 6)

End-of-19th century BRs can be classified into two categories: very short ones that were a minority and very long ones that, by far, formed the great majority. The very short ones (no more than 5-6 lines) were merely descriptive. As a matter-of-fact, the shorter, the less critical.

These very short 19th century BRs remind us of those published by the end of the 18th century which were mere summaries of the book content, i.e., BR where the book reviewer's opinion had no place at all. Let us see what Andrew Duncan (the then anonymous editor of *The Philosophical and Medical Commentaries*, the first English-medium journal that published medical BRs) had to say regarding the role of the book reviewer in the closing years of the 18th century:

> As it is not our intention to offer any opinion with regard to the general characters of the books, we shall, on every occasion avoid, as much as possible, either applauding or condemning any author. Our chief aim will be, to give such a view of books as may enable every reader to judge for himself. Whether the writers ought to add to this their own judgments, is with us a doubtful point. If their account be sufficiently accurate and full, its seems to supersede the necessity of any addition of their own.

As I said before, the great majority of end-of-19th century BRs were very long (15 pages or more). What is more, most of them were harsh, personal and very detailed critical remarks to the book content.

3.1.2. The judge's voice: direct and personal (Table 1, row 7)

How were these critical remarks voiced? As the following examples clearly illustrate, they were very personal (i.e., voiced at the book author) and very direct:

(1) The section on distortions by Dr. James contains more inaccuracies than any article which has been published on the subject.

(2) The burden of responsibility for the countless inconsistencies in the entire work must rest entirely on Dr. Leeds.

(3) We are at a total loss to understand why Dr. Wilkinson uses such an outdated terminology.

Another characteristic feature of end-of-19th century BRs consists in the fact that book reviewers tended to end their reviews with a global, harsh criticism, such as:

(4) Starr's manual, as a whole, is unsatisfactory.

(5) McMurtry's monograph contains nothing which should be new to the specialist.

3.1.3. Criticism targets: medical praxis (Table 1, row 8)

Treatment methods and surgical procedures were quite frequently the targets of criticism, especially when comparing the way these methods/procedures were performed on both sides of the Atlantic (USA vs. UK). The reviewer would then go at great length to explain why, how and to what extent he disagreed with the book author. Criticisms were also targeted at questions of nomenclature, definition or classification.

However, the closer we get towards the year 1900, the more important do visual elements become: table presentation, illustrations and drawings were more and more frequently the targets of critical remarks. This means that it is from the end of the 19th century that scientists began to integrate the visual and the textual and to exploit the cognitive possibilities of visual elements.

3.1.4. The judge and the accused (Table 1, rows 9-10)

The previous examples allow us to observe two quite interesting things: the book reviewers' presence in the BR itself, on the one hand, and, on the other, the identification of the entity responsible for the flaws put forwards in the BR.

In examples 1 to 5, the pronoun *we* and *us* identify the book reviewer. These pronouns cannot refer to several persons since, as I said before, back then, BRs were written by one reviewer only[3]. We could

3 Even today it is rare for BRs to be written by several reviewers.

thus speculate that *we/us* were used as the voice of experts talking on behalf of the entire scientific community.[4]

As regards the accused (i.e., the entity held accountable for the flaws observed in the book), examples 1 to 5 above show that it was the book author, clearly identified by his name and surname, who was responsible.

3.2. Mid-20th century book reviews (1950–1960)

3.2.1. General features

a) Type of books, number of authors and their identification (Table 1, rows 2 and 3)

As Graph 1 shows, although authored books (64%) still outnumber edited books (30%), the latter increased considerably in comparison with the previous period. Moreover, in both types of books the number of authors (now only identified by their names and surnames) has increased sharply. By contrast, single-authored books have significantly decreased in comparison with the previous period.

b) Languages of publication (Table 1, row 5)

As Graph 3 shows, all the books reviewed but two were written in English.

c) Book review length and identification of the book reviewer (Table 1, rows 4 and 6)

The book reviews are now shorter, and the reviewer is always identified by his name and surname followed by his academic status. Let us now analyse the reviewer's critical voice (Table 1, row 7).

---

4 The use of these two pronouns could also be a *calque* from French rhetoric that hates the arrogant-sounding *Je/I*. Let us keep in mind that French was an important language of scientific knowledge dissemination by the end of the 19th century.

3.2.2. The judge's voice: not as verbose, but still very polemical, aggressive and passionate

The judge's tone of voice is very similar to that observed in end-of-19th century BRs but it is more arrogant, face-threatening and devastating, as examples 6 to 14 below illustrate:

(6) It is an absolute heresy to assert, as Swartz does, that diphtheria is of little or no importance in West Africa.

(7) Trauma and Disease is a very bad book. It amounts to such a blatant attack on intelligent inquiry. It portrays the worst form of mental terrorism used against any who is interested in the facts.

(8) For me, the book is very disappointing. That the total number of pages remains the same (as that of the original description of the disease) is due to arbitrariness on the author's part. The senior author fails to discuss important issues. He completely fails to deal with the fundamentals of this subject.

It was frequent too for the reviewer not only to write a globally negative BR, but also to end his review with a harshly expressed criticism:

(9) My advice to the would-be-buyer who is about to hock a prized belonging to raise $60 – the book sky-high price – is very simple: Don't.

(10) Anyone buying the book on the basis of the title and cover is in for a disappointment.

The sarcastic and cynical tone of the final verdict is quite evident in the following example:

(11) Psychiatrists and psychologists ought to read the book, even if only to be enraged.

Contrary to what is so frequently observed in today's BRs (see below), in mid-20th century BRs, the reviewer did not resort to 'sugar-coating' expressions to sweeten his critical remarks.

A rhetorico-pragmatic feature of mid-20th century BRs was the sarcastic/humorous tone of their criticisms. The following example was drawn from the review of the book entitled *Physical Diagnosis*, published in 1958 right in the middle of the cold war. The book reviewer suggests the adoption of a brand new approach to physical diagnosis in the following terms:

(12) I hereby contribute the following suggestions gratis. Got the idea? OK. Your 'text' should consist of some records of sounds, murmurs, etc. to replace your totally inadequate written words describing such phenomena. Finally, your 'text' should include a 10 to 15 minute film showing an expert diagnostician performing a complete physical examination, from beginning to end. (For this last section, I recommend that the editors of Playboy be canvassed to get nominations for the examinee.) All right, so snicker. I'll bet the Russians are doing it already.

Another book reviewer ends his review in a very sarcastic way (and after a very long list of negative remarks about the book):

(13) After overlooking all the defects (which is very hard, alas!)

Full of humour, too, is example 14:

(14) Eysenck quotes on page 2 what must be the longest word in the English language (antifloccinaucinihilipilificationistically).

3.2.3. Targets of criticisms (Table 1, row 8)

a) Conceptual criticism

Medical praxis is not a target of criticism any more. In the mid 20th century, books are rather criticised for lacking information and for the fact that bibliographical references are trivial, erroneous and/or incomplete (examples 15 and 16 below).

(15) The author admits in the Preface that references have been included without verification… Besides, he cites trivial or discredited contributions with reckless eclecticism.

(16) Whatever possessed the authors to lump together almost 1,000 references contributed by 45 authors and covering a multitude of sometimes not so closely related subjects in one alphabetical list will remain a mystery forever. As it stands now, the list of references is useless and will defy all but the fiercest and most persistent investigators.

b) Contextual criticisms that reveal social concerns.

The book price is very frequently a target of criticism in mid-20th century BRs (see example 9 above). Such criticism is not to be found in any BR of either one of the two other periods considered here.

The concept of audience is often found as a critical remark, as in example 17 below:

(17) For whom is this expensive, Gargantuan, grotesque opus intended? The essence of a good textbook is the judicious omission of the irrelevant. This book exemplifies the opposite approach, which may be termed 'from soup to nuts'.

c) Errors (typographical and interpretative)

Stylistic errors (e.g., grammar, spelling) were quite frequently mentioned as well as errors of interpretation in tables and/or graphs. This last type of criticism is interesting in that it evinces the growing importance of the visual in scientific argumentation. Such errors are qualified as "too many", "an inexcusable collection", "too numerous to be counted", "countless" and "glaring".

3.2.4. The judge and the accused (Table 1, rows 9 and 10)

The book reviewer now identifies himself with the first personal pronoun *I* or with the expression *My opinion*. In other words, it is not the expert talking on behalf of the scientific community as it used to be in the previous period, but it is a "common reader", a reader as any other. This use of the first personal pronoun helps toning down the criticism. In one BR only did we find the expression *This reviewer*.

Now, who is responsible for the flaws mentioned in the BR? As in the previous period, it is the book author, but he now appears as an indefinite person. It is not Walter Smith, let's say, who is held accountable, but it is simply the author. Much less frequently the book stands on the bench of the accused.

3.3. Beginning of the 21st century (2000–2005)

3.3.1. General features

a) Type of books and number of authors (Table 1, rows 1-2)

From the 1990's on, the book title is followed by the publisher's phone number, and in the year 2005, the publisher's web page started making its appearance. These are interesting additional details that mirror the fact that today's science is big business (see below).

Moreover, as can be observed from Graphs 1 and 2, the number of co-edited book increased significantly as well as the number of authors both in authored and in edited works.

b) Book authors' and reviewers' identification (Table, rows 3 and 4)

Book authors are now identified with their first name initials and their surnames. Nothing else. By contrast, reviewers are identified with their names and surnames, their academic degree, medical specialty and workplace, i.e., with many more details than ever before and many more details than the book authors themselves, as if the book reviewer were more important than the book author.

c) Languages of publication (Table 1, row 5)

Graph 3 clearly shows that all the books reviewed are now originally published in English.

d) Book review length (Table 1, row 6)

As the Table indicates, it is quite variable, but the standard deviation is much smaller than it was in the previous periods.

3.3.2. The judge's voice: neutral and conciliatory (Table 1, row 7)

In the BRs published at the beginning of the 21st century, critical remarks are still quite directly voiced but in a much less emotional and face-threatening fashion than in the two previous periods.

Furthermore, and contrary to what was observed before, negative comments are almost always followed by positive remarks that are generally preceded by contrastive connectors, such as *however* or *nevertheless* (cf. example 18):

(18) A major weakness of the book is its lack of depth in some subject areas.... However, this text is an excellent collection of articles written by outstanding expert authors.

In other words, today's BRs ("today" means from the 1990's on) never end on a negative comment. It is interesting to note that such a final 'good news move' seems to be a distinctive feature of today's BRs from a wide array of disciplines (cf. Motta Roth 1998 and Hyland 2000).

As for humour, we found one example only (see example 19 below) of (black) humour in the 21st century sample of BRs. It is interesting to note that the book author is moreover clearly identified by his surname as if the book reviewer wanted to be even more destructive.

(19) My reaction to Singer's book is akin to discovering that a friend has served me her pet for dinner.

3.3.3. Targets of criticisms (Table 1, row 8)

The most frequent criticism refers today to redundancy between chapters, poor chapter integration, lack of 'balance' of the work under review (viz., some chapters present a deep and thoughtful analysis of a given issue, while others are rather superficial), lack of cohesion, and arbitrary and superficial editing. This type of criticism is due to the fact that, as we saw before, the great majority of the books reviewed in this period are edited books written by various contributors. We found a book, for example, that was written by 121 authors! Reaching a stylistic and conceptual balance with 121 authors is an arduous task indeed, because each contributor has his/her own style, level of expertise and competence.

Secondly, when bibliographical references are the target of criticism, it is because they are obsolete, not because they are incomplete or inaccurate, as was the case in the mid-20th century. It thus seems that reference recency is today a very important criterion to judge the worth and quality of an academic book.

Regarding errors (web page and statistics), errors mentioned in today's BRs are no longer related to either orthography or typography not only because all computers are equipped with a built-in dictionary, but also because all publishing houses have their own linguistic editors. Errors rather indicate wrong web pages and/or incorrect use of statistics. This once more reflects the importance of quantification.

3.3.4. The judge and the accused (Table 1, rows 9 and 10)

The judge's voice hardly makes itself heard in today's BRs and when it does, it is by means of the *I* personal pronoun (e.g., *I would have preferred, I cannot agree with*) or the possessive adjective followed by a noun (*My only concern in this respect).* Such expressions reduce the force of the criticism because the judge now plays the role of an ordinary reader, not that of an expert[5].

What about the accused? It is not the book author any more who is held accountable for the flaws mentioned in the BR but the book itself or a book chapter. In other words, it is an inanimate object (not a person) that stands on the bench of the accused. This is what is called a total objectivisation or impersonalisation of critical speech acts.

4. Discussion: what has changed and why?

The aforesaid evinces the increased standardisation, hyperspecialisation and professionalisation of science as well as the increasing social and commercial character of the scientific enterprise. Let us analyse the arguments that support such an assertion.

5 It is worthwhile mentioning that this progressive objectivisation/impersonalisation of critical speech acts has been observed as well in other academic genres, but with other rhetorical strategies (Bazerman 1988; Swales 1990; Kuo 1999).

4.1. Knowledge dissemination

We have seen that end-of-19th century BRs are much longer (about four times as long) than later ones. Two factors explain the difference observed. The first factor is rhetorical: In the 19th century, the book reviewer, almost always anonymous, used to write long verbatim quotes from the book he was reviewing in order to explain, with great details, how and why his opinion differed from that of the book author. Moreover, he used to make long digressions about topics that were not directly related to the uttered criticisms. We have to keep in mind that there was no space/word limitation at that time; as a consequence, the book reviewer felt free to be as verbose as he wanted.

The second factor is historical-conceptual: indeed, over time books lost their importance as primary means of knowledge dissemination and were replaced by research papers, especially from the beginning of the 20th century (Salager-Meyer 1998). This explains why today's BRs are not only much shorter than they used to be, but also why they appear in the last section of medical journals.

But it is not only in the medical field that books became secondary in the dissemination of knowledge. Swales (1990), for example, reports that since the 1960's references to books have significantly diminished in the papers published in *TESOL Quarterly*. However, it is worthwhile mentioning that, according to the *Social Science Index*, 70% of prestigious linguistics journals publish BRs vs. 40% in economics journals and 35% in chemistry (Giannoni 2002: 336). These figures clearly show that, in spite of the fact that BRs are today less important than research papers in scientific knowledge dissemination, their importance differs from one discipline to another.

From a purely business perspective, journal space costs a lot of money to editorial houses and, because since the 1960's the number of scientific papers being published has proliferated at a speed "as startling as the overnight increase of bacteria in a petri dish" (Gross *et al.* 2002: 29)[6], journals must leave more space to what today represents scientists'

---

6 According to de Solla Price (1986), there were less than ten scientific journals in the 17th century, about 100 in 1800, and 10,000 in 1900. This, of course, depends on how define what a 'scientific journal' is, but it is undeniable that the explosion of scientific journals started in the closing years of the 19th century.

'bread and butter' or 'staple food': the sacrosanct research paper. This is why, in order to save space, Anglo-American pedigree medical journals now publish what they call 'book notes' which remind us a lot of the 'structured abstract' (see Appendix for an example of such a book note). It is interesting to remark that one of the nine obligatory moves of the 'book note' is labelled 'Limitations', a move where the reviewer states his/her criticisms.

4.2. From the authored book to the edited book

Traditionally, books were written by a single author, but the increase in the number of scientists and in the complexity and/or hyper-specialisation of scientific research (especially in the natural and hard sciences) progressively led to an increase in the number of books written by several authors and in the number of edited books. This "multiple authorship" or "mega-authorship" phenomenon (Kretschmer/Rousseau 2001: 612), also called "hyperauthorship" (Cronin 2002: 560), "mass recognition" (Morris *et al.* 2005) or "polyauthoritis giftosa" (Modi *et al.* 2008: 6), is very well known in scientific research papers (Glänzel/Schubert 2004; National Science Board 2004; Weeks *et al.* 2004; Cronin 2005; Meadows 2005; Durani *et al.* 2007).

One of the consequences of the increasing collaboration among scientists that Gross *et al.* (2002: 233) refers to as "colaboratories" or "invisible colleagues" is that authored books are becoming a historical curiosity as is the single-authored research paper and not only in the medical field but in others as well. As Cronin (2005) claims, "the lone wolf scholar" no longer exists, except in disciplines such as philosophy where books still are the "gold standard for tenure and promotion in the humanities" (Cronin/La Barre 2004: 92), at least in North-American universities.

But there is another factor that explains why co-edited books today outnumber single-authored ones. It has to do with the scientific research evaluation system, a system that is very similar all around the world, whether it be in developed or in developing countries and is characterised by being extremely competitive and obsessed with numbers, with quantity in detriment of quality: impact factor, h-index and citation counts are the Holy Trinity of scientific production markers.[7] In this respect, a senior researcher from the behavioural sciences, interviewed in Anderson *et al.*'s study (2007: 440) on the perverse effects of competition on scientists' work and relationships, refers to university productivity in the following terms:

> […] numbers mean everything. you know, the quality of my work [...] they don't even read your papers. In my field, all they care about is how many papers you publish per year. They don't care whether it's an excellent study or a study that people don't really believe.

> Over the past 30 years, academic science in the USA has come to resemble more closely its counterpart in the corporate sector. During this period, science has increasingly come to reflect the assumptions, tools, methods and products of modern day market capitalism, a political and economic system driven by competition.

4.3. Targets of criticisms

We have seen that the targets of criticisms have changed a lot over time. In the 19th century, medical praxis constituted the main target of the book reviewer criticisms; then it was the book author who was blamed for having omitted important information or for having misquoted bibliographical references and, at the beginning of the 21st century, it is the book or a book chapter that is criticised for being redundant or lacking cohesion.

Moreover, the increasing number of criticisms made to experimental/statistical methods clearly evinces a fundamental change in the scientific method, a hallmark for understanding the structure of

---

7 The impact factor is now a widely accepted criterion to assess the quality of scientific journals (cf. Dong *et al.*, 2005). But it has been recently debated and criticised. Egbert (2007), for example, warns against the use of the impact factor as the sole indicator to determine the value of a given journal. Sharon Begley (2006), for her part, denounces the fact that a standard procedure of some leading journals is to send an email to all submitting authors, asking them to include citations from these journals to boost their impact factor. The executive director of the American Physiological Society, which publishes 14 journals, put it even more bluntly by asserting that "We have become the whores of the impact factor" (cited in Begley 2006).

past and present scientific thought. Indeed, we observed a switch from a structure essentially based on mere observation to a critical structure based on experimental results that convert into scientific facts and constitute today's paradigm of scientific argument. As Poovey (1998: 56) expresses: "Numbers have come to epitomize the modern fact".

Along the same lines, the increasing number of criticisms uttered at visual elements (tables, figures, graphs, etc.) mirrors the fact that the interaction between the visual and the textual is fundamental in the construction of today's scientific argumentation.

4.4. Social and commercial character of today's science

The frequent criticisms voiced at the unrecency of bibliographical references in the BRs published at the beginning of the 21st century, and the two obligatory moves called 'Related reading' and 'Audience' in the highly structured 'book notes' mirror, on the one hand, that the scientific argument is more and more contextualised in an existing body of knowledge, and, on the other, that science has increasingly become an eminently social practice.

We also observed that over time science became a commercial activity: it is only in the last year of the 20th century that BRs started indicating the book price, and, a century later, it is followed by the publishing house web page in order to facilitate the rapid purchase of the book. This is undoubtedly very useful information that reflects the prevailing 'spirit of the time': business, efficiency, rapidity, maximum profit.

As an early career academic from the biomedical sciences reported in Anderson's *et al.*'s study (2007: 458) claims:

> Science is business and so it is just all money. At one level nobody cares what you are doing [...] The whole intellectual content of ... the whole idea of science as intellectual inquiry becomes very secondary.

He further adds:

> We've changed this university from the production of science ... to the production of dollars. And the need to create dollars [...] has created the subculture of science, overshadowed by the larger issue of entrepreneurial, avaricious, crave dollars.

4.5. *Increasing scientific monolingualism*

The fact that 100% of the books reviewed in the last period studied are written in English is a clear reflection of the now well-documented increasing monolingualism in scientific communication (cf. Carli/ Ammon 2008, and the numerous voices that defend linguistic diversity)[8].

4.6. *The judge's social concern*

As we have seen before, over 60% of the BRs published at the end of the 19th century were anonymous or only mentioned the book reviewer's first name and surname initials. By contrast, all the BRs from the mid-20th century and beginning of the 21st century are not only carefully signed but also indicate the reviewer's place of work and medical specialty.

As regards the reviewer, it was first a polemical, arrogant, aggressive, passionate, sarcastic and threatening character who used to voice his harsh criticisms to a clearly identified individual (the book author). This judge became more and more neutral and detached and started voicing his unemotional critical remarks to an inanimate object (the book or a book chapter). Yesterday's judge was that of a polemical 'science in the making'; today's is that of a cooperative, humble, respectful and harmonious 'science once-made'.[9] It is as though today's book

8 This linguistic 'evolution' goes hand in hand with that observed by Gross *et al.* (2002: 195) in scientific research papers written in French and published in *Le Journal de Physique*. Indeed, at the end of the 19th and beginning of the 20th century, bibliographical references in these articles referred to papers written in French and German. It is only from the 1970's on when the references started being made to papers written in English. The same evolution was observed by Navarro / Alcaraz Ariza (1997) in medical papers published in Spanish-medium journals.

9 It is interesting to note that the destructive and aggressive tone of critical comments has not totally disappeared from written medical discourse. It is to be found, indeed, in the academic genre 'Letters to the editor', especially when the topics being discussed refer to hot issues, such as the war between "alternative/complementary medicine" (CAM) and "scientific/conventional/

reviewer were trying to minimise the interpersonal damage of his critical comments, to negotiate the social relations inherent to the BR genre and to establish a solidarity framework with the book's author. Such concerns were totally foreign to the 19th and even mid-20th century judge.

5. Conclusion

The changes observed in the general features of the BRs and in the rhetorical evolution of criticism reflect changes not only in the scientific enterprise but also in the scientific society in general. Such changes should not be considered as a sign of progress or improvement but as a process of selection and adaptation to the increasing volume of scientific papers, to the needs and the increasing complexity of the context in which scientific activity develops and to the changes that, as a consequence, are suffered by the scientific enterprise whose actors – who come from different linguistic and geographical horizons – must struggle to make themselves visible on the Big Science stage.

Has the BR reached its generic (structural, stylistic and rhetorical) maturity? I do not think so. We can be quite sure that the BR (as well as the research paper) will keep evolving as a response to the requirements of disciplines, scientific societies and new communication technologies. As a matter of fact, I strongly believe that BR will keep evolving as long as there will be books to review. The highly structured 'book note' very well illustrates my point.

I hope that this paper has shown, on the one hand, how enriching and helpful diachronic studies of scientific discourse can be for a better

---

orthodox" medicine. I recently found an excellent example of such devastating language in a letter to the Editor entitled *"Complementary medicine: a very bad report on regulating complementary medicine"* (Colquhoun 2008), published in *the British Medical Journal* in July 2008, in which the author pejoratively refers to CAM as "magic medicine", "theatrical placebo" and "mumbo-jumbo". Moreover, Dr. Colquhoun qualifies Pittilo's report (a report that assesses CAM efficiency) as "execrably bad". The tone of such criticisms clearly reminds us of that observed in BRs published in the closing years of the 19th century.

understanding of the formal and rhetorical features of today's science and, on the other, that today's discourse can only be properly understood by making reference to its historical evolution.

References

Anderson, Melissa S. / Ronning, Emily A. / De Vries, Raymond / Martison, Brian 2007. The Perverse Effects of Competition on Scientists' Work and Relationships. *Science and Engineering Ethics*. 13/4, 437-461.

Atkinson, Dwight 1996. The Philosophical Transactions of the Royal Society of London, 1575-1975: A Socio-historical Discourse Analysis. *Language in Society*. 25, 333-371.

Bazerman, Charles 1988. *Shaping Written Knowledge*. Madison. University of Wisconsin Press.

Begley, S. 2006. Scientific Journals Artfully Try to Boost their Ratings. *The Wall Street Journal*. June 2006.

Bouveresse, Jacques 1984. *Le Philosophe chez les Autophages*. Paris: Editions de Minuit.

Brinton, Laurel J. 2001. Historical Discourse Analysis. In Schiffrin, Deborah / Tannen, Deborah / Hamilton, Heidi (eds) *The Handbook of Discourse Analysis*. Malden, Massachusetts & Oxford: Blackwell, 138-160.

Carli, Augusto / Ammon, Ulrich 2008. Linguistic Inequality in Scientific Communication Today. *AILA Review*. 20.

Colquhoun, David 2008. Complementary Medicine: A Very Bad Report on Regulating Complementary Medicine. *British Medical Journal*. 337, a591.

Cronin, Blaise 2002. Hyperauthorship: A Postmodern Perversion or Evidence of a Structural Shift in Scholarly Communication Practices? *Journal of the American Society for Information Science*. 52/7, 558-569.

Cronin, Blaise 2005. *The Hand of Science*. Lanham, MD: Scarecrow Press.

Cronin, Blaise / La Barre, Kathryn 2004. Mickey Mouse and Milton: Book Publishing in the Humanities. *Learned Publishing*. 17/2, 85-98.

De Solla Price, Derk J. 1986. *Little Science, Big Science and Beyond.* New York: Columbia University Press.

Dong, Peng / Loh, Marie / Mondry, Adrian 2005. The Impact Factor Revisited. *Biomedical Digital Libraries*. 2, 7. <http://www.bio-diglib.com/content/2/1/7>

Durani, P. / Rimouche, S. / Ross, G. 2007. How Many Plastic Surgeons Does It Take to Write a Research Article? Authorship Proliferation in and Internationalization of the Plastic Surgery Literature. *Journal of Plastic Reconstruction and Aesthetic Surgery.* 0, 956-957.

Glänzel, Wolfgang / András, Schubert 2004. Analysing Scientific Networks through Co-Authorship. In H.F. Moed *et al.* (eds) *Handbook of Quantitative Science and Technology Research*. Kluwer Academic Publishers, 257-276.

Gross, Alan G. / Harmon, Joseph E. / Reidy, Michael 2002. *Communicating Science. The Scientific Article from the 17th Century to the Present*. Oxford: Oxford University Press.

Egbert, Joy 2007. Quality Analysis of Journals in TESOL and Applied Linguistics. *TESOL Quarterly*. 1/41, 157-171.

Fleischman, Suzanne 1990. Philology, Linguistics, and the Discourse of the Medieval Text. *Speculum*. 65, 19-37.

Fries, U. 1983. Diachronic Textlinguistics. In Hattori, Shiro / Inoue, Kazuko (eds) *Proceedings of the XIIIth International Congress of Linguistics*. August 29-September 4, 1982. Tokyo: Tokyo Press, 1013-1015.

Giannoni, Davide Simone 2002. Hard Words, Soft Technology. Criticism and Endorsement in the Software Review Genre. In Gotti, Maurizio / Heller, Dorothee / Dossena, Marina (eds) *Conflict and Negotiation in Specialized Texts*. Bern: Peter Lang, 335-363.

de Granda, Germán 1980. Historia social e historia lingüística en Hispanoamérica. In *Simposio Internacional de Lengua y Literatura Hispánicas*. Bahía Blanca: Universidad Nacional del Sur, 203-215.

Hyland, Ken 2000. *Disciplinary Discourses: Social Interactions in Academic Writing*. London: Longman.

Jucker, Andreas H. (ed.) 1995. *Historical Pragmatics. Pragmatic Developments in the History of English*. Amsterdam: John Benjamins.

Kretschmer, Hildrun / Rousseau, Ronald 2001. Author Inflation Leads to a Breakdown of Lotka's Law. *Journal of the American Society for Information Science and Technology*. 52/8, 610-614.

Kuo, Sai-Hua H. 1999. The Use of Personal Pronouns: Role Relationships in Scientific Journal Articles. *English for Specific Purposes*. 18/2, 121-138.

Meadows, Jack 2005. Differences between the Sciences in their Handling of the Research Literature. *European Science Editing*, 31/4, 113-116.

Milroy, James 1991. *Linguistic Variation and Change: On the Historical Sociolinguistics of English*. Oxford: Blackwell.

Modi, Paul / Hassan, Ansar / Teng, Carolyn J. / Chitwood, Walter Randolph 2008. How Many Cardiac Surgeons Does It Take To Write A Research Article? Seventy Years of Authorship Proliferation and Internationalization in the Cardiothoracic Surgical Literature. *The Journal of Thoracic and Cardiovascular Surgery*, 136, 4-6.

Morris, Steven, A. / Golstein, Michel L. 2005. Manifestation of Research Teams in Journal Literature: A Growth Model of Papers, Authors, Collaboration, Co-Authorship, Weak Ties and Lotka's Law. *Journal of the American Society for Information Science and Technology,* 58/12, 1764-1782.

Motta-Roth, Désirée 1998. Discourse Analysis and Academic Book Reviews: A Study of Text And Disciplinary Cultures. In Fortanet, Inmaculada / Posteguillo, Santiago / Palmer, Juan Carlos / Coll, Juan Francisco (eds) *Genre Studies in English for Academic Purposes.* Castellón de la Plana: Universitat Jaume I, 29-59.

National Science Board 2004. Science and Engineering Indicators. <http://www.nsf.gov/sbe/srs/seind04/>

Navarro, Fernando A. / Alcaraz Ariza, María Ángeles 1997. El idioma de la dermatología en España a través de las referencias bibliográficas publicadas en Actas Dermo-Sifiliográficas entre 1910 y 1995. *Actas Dermosifiliogr*. 88, 358-364.

O'Brien, Eoin 1980. The Lancet Maketh the Man? Sir Dominic John Corrigan (1802-80) *The Lancet.* 1356-1357.

Poovey, Mary 1998. *A History of the Modern Fact: Problems of Knowledge in the Sciences of Wealth and Society*. Chicago: University of Chicago Press.

Romaine, Suzanne 1982. Socio-historical Linguistics. Cambridge: Cambridge University Press.

Salager-Meyer, Françoise 1998. Books vs. Journals: A Diachronic Study of Referencing in Written Medical English Prose (1810–1995). *The ESPecialist.* 18/2, 147–185.

Salager-Meyer, Françoise / Alcaraz Ariza, María Angeles 2004. Negative Appraisals in Academic Book Reviews: A Cross-Linguistic Approach. In Candlin, Chris / Gotti, Maurizio (eds) *Intercultural Aspects of Specialized Communication.* Bern: Peter Lang, 149-172.

Swales, John 1990. *Genre analysis. English in Academic and Research Settings*. Cambridge: Cambridge University Press.

Valle, Ellen 1997. A Scientific Community and its Texts: A Historical Discourse Study. In Gunnarson, Britt Louise / Linell, Per / Nordberg, Bengt (eds) *The Construction of Professional Discourse.* London. Longman, 76-98.

Weeks, William B. / Wallace, Amy E. / Surrot, Kimberly B. C. 2004. Changes in Authorship Patterns in Prestigious US Medical Journals. *Social Science and Medicine.* 59, 1949-1954.

Appendix

Field of medicine	Quality improvement and patient safety.
Format	Soft-cover book.
Audience	Quality improvement leaders, physicians, hospital executives, managers
Purpose	To provide an overview of …
Content	Eight chapters cover… Each chapter includes.. Some also include…
Highlights	The book offers a useful update on …The material on … may be especially...
Limitations	This collection lacks a unifying introduction… Many of the articles are self-promoting… Many recent initiatives on quality management are never mentioned… Much of the information is available from any medical library or on the Internet.
Related readings	Several websites provide information about… including http://www……
Reviewer	Patrick S. Romano, MD. MPH, University of California. Davies School of Medicine. Sacramento. California.

Table 2. Example of a 'book note'. Review of *Medical Quality Management Sourcebook*. 2004 edition.

Luz Gil-Salom / Carmen Soler-Monreal

Appraisal Resources in Scientific Research Article Discussions

1. Introduction

Research on academic writing has covered a wide range of topics. Genre-based studies have mainly explored the organisational patterns of research articles (hereafter RAs) on the basis of the concepts of *move* and *step* proposed by Swales (1990) to describe the *CARS* (*Create a Research Space*) model of introductions (Dudley-Evans 1994; Posteguillo 1999; Lewin *et al.* 2001; Peacock 2002; Samraj 2002; Ruiying/Allison 2003, 2004; Soler-Monreal *et al.* 2006a, 2006b; Soler-Monreal/Gil-Salom forthcoming). Other studies have focused on linguistic signals and semantic features expressing attitude and opinion, particularly the writer's strategies to combine factual information with personal judgement and interaction with the reader (Latour/Woolgar 1979; Hunston 1994, 2003; Salager-Meyer 1994; Skelton 1997; Hyland 1998a, 1998b, 1998c, 2001, 2002). To refer to the language of opinion and personal feelings of the writer, different labels have been used: evaluation (Hunston 1994, 2003), stance (Conrad/Biber 2003) and appraisal (Martin 2003). In addition, the writer's judgements have been related either to attitudinal meanings and value systems or to degrees of likelihood, evidentiality (Chafe/Nichols 1986) and modality (Halliday 1994).

Several different taxonomies have been proposed to classify the set of meanings and related resources which explicitly express particular subjective positions. According to Hunston (1994: 192), evaluation involves a statement of personal judgement and an appeal to shared norms and values. Linguistic resources of evaluation indicate the writer's attitude to the content of a text in terms of status in a *certain-uncertain* scale of evaluation, of value in a *good-bad* scale of evaluation, and of relevance of an entity in an *important-unimportant* scale of evaluation. On the other hand, Conrad/Biber (2003) distinguish between epistemic, attitudinal and style stance. Epistemic stance comprises comments on certainty, reliability and limitations of a proposition. Attitudinal stance conveys the speaker's attitudes, feelings or value judgements. As for style stance, it describes the manner in which the information is being presented. A more extensive classification is provided by Appraisal theory (http://www.grammatics.com/appraisal), an ongoing research project which explores the importance of distributing interpersonal meanings that enable the development of the writer's point of view across a text. It includes evaluation and attitudinal stance in the attitude semantic domain, while epistemic stance, evidentiality, epistemic modality and hedging are considered to fall under the engagement appraisal subsystem.

Research on lexico-grammatical means of imposing attitudes, interpretations and assessment of truth value on readers has been carried out in genres related to the RA in specific disciplines (Thompson/Ye 1991; Hyland 1998a, 1998b, 1998c, 2001; Gil-Salom *et al.* 2001; Stotesbury 2003; Koutsantoni 2004). Based on a corpus of RAs, Koutsantoni's (2004) classification comprises attitude, certainty and common knowledge markers. Whereas attitude markers include devices that express writers' affective values towards propositional contents and/or readers, certainty markers emphasise authors' certainty and conviction to claims and ideas. On the other hand, common knowledge markers explicitly invoke received knowledge and take certain propositions for granted. Hyland (1998b) uses the cover terms *hedges* and *boosters* to refer to expressions conveying both epistemic and affective meanings. Their role is to adjust the truth-value or certainty attached to a statement and to maintain relations between writer and reader by assuming shared attitudes, values and reactions, as well as group membership.

Studies concerned with the resources mentioned above have been conducted along whole texts. However, to the best of our knowledge, there has been no research exclusively focused on the types of subjective language used by researchers in specific sections of RAs. Our interest in this study has been to explore some appraisal resources in the discussion sections of engineering research articles. The main communicative purpose of the discussion section is to persuade the academic community of the validity of the research presented and the convenience of accepting new knowledge claims (Latour/Woolgar 1979; Hunston 1994; Ruiying/Allison 2003). In discussion sections, writers comment on the data presented in the results section and make their own views on the study explicit. Findings are highlighted and situated in the context of a wider body of knowledge by relating results to the work of others with the aim of establishing credibility so as to consolidate the writer's research space. This results in a number of implicit and explicit realisations of personal judgement, commitment to the truth-value of propositions and attitude towards the audience.

All this considered, the purpose of the present paper is to describe evaluative, certainty and epistemic expressions in the discussion sections of engineering texts.

2. Corpus and method

The study is based on a corpus of RAs published between 2001 and 2003 in 18 specialised leading journals in the disciplines of Computing, Telecommunications, Nanotechnology and Robotics. Drawing on the assumption that the discussion section concentrates the largest number of appraisal resources in the RA, focus was placed on the closing sections which contained the word *discussion* in their headings. Forty-six discussion sections were selected: 11 from Computing, 10 from Nanotechnology, 15 from Robotics and 10 from Telecommunications.

The study was carried out in three phases. First, a manual and intuitive genre-based approach provided a broad spectrum of lexical, grammatical and discourse resources expressing the communicative purposes of the discussion texts in the corpus. Second, the texts were searched using *TextWorks* (Gil-Salom *et al.* 2004), a corpus analysis tool which displayed the number of occurrences and the concordances of the most frequently used items showing the writer's opinions and feelings. Third, the presence of these items in the move-step structure of the discussion sections of our corpus was checked.

Here, we focus exclusively on the analysis of attitudinal, certainty and epistemic adjectives and adverbs. Other lexical and grammatical devices, such as nouns and modals, can also be used to mark the expression of personal feelings and assessments but are beyond the scope of this paper.

3. Results

The taxonomies suggested by Koutsantoni (2004) and Hyland (1998a, 1998b, 1998c, 2002) have been at the origin of the classification presented here, and have helped us to recognise evaluative, certainty and epistemic lexis. Some of the items analysed coincide with those pointed out in their taxonomies; others have been identified in the corpus of RAs under study. The number of items of each category is shown in separate tables. The communicative purpose of the statements within which these items are used is taken into account, and reference is made to the moves and steps of the discussion sections where these items appear. The move-step structure is based on Ruiying and Allison's (2003) two-level macrostructure account for RA discussions (see Appendix, Table 9).

3.1. Expressing attitude

Koutsantoni (2004) classifies evaluative adjectives and adverbs, intensifying adverbs and attitudinal stance adverbs as attitude markers. Appraisal theory includes evaluative adjectives and adverbs and intensifying adverbs in the appreciation subsystem of attitude as they indicate evaluation in terms of social value. In contrast, attitudinal stance adverbs belong to the affect subsystem of attitude, since they establish interpersonal bonds with readers and guide them in the acceptance of their claims.

Evaluative adjectives and adverbs

Explicitly evaluative adjectives form the most common group of evaluative resources in the discussion sections of our corpus. Table 1 provides the quantification of evaluative adjectives present in the corpus. Of these, five adjectives clearly predominate over the rest: *significant* (75 occurrences), *important* (73 occurrences), *effective* (37 occurrences), *critical* (30 occurrences) and *interesting* (29 occurrences). There are differences, however, in the distribution of these adjectives across disciplines. Although *significant* is the most common adjective used in the field of Computing, Robotics prefers *important.* The figures in Nanotechnology, on the other hand, seem to indicate that *significant, important, effective* and *critical* are almost equally used for assessment. In comparison, the low figures in Telecommunications suggest that this area recurs to other grammatical categories to express attitude and appreciation.

	Comp[1]	Telecom	Nanotech	Robot	Totals
significant	27	5	33	10	75
important	15	9	29	20	73
effective	5	5	24	3	37
critical	1	1	21	7	30
interesting	4	3	8	14	29
advantageous	2	1	1	0	4
acceptable	0	1	0	3	4
valuable	0	3	0	0	3
attractive	0	1	0	0	1
ill-suited	0	1	0	0	1

Table 1. Number of occurrences of evaluative adjectives.

Significant is particularly common in comments on results (Move 4), where the researcher speculates on the data obtained and evaluates

1 Abbreviations are as follows: Comp for Computing; Telecom for Telecommunications; Nanotech for Nanotechnology; and Robot for Robotics.

the findings with the purpose of occupying a niche in the discipline community. Arguments combine evidence with personal attitudes and opinions which propitiate interpersonal relationships and predispose readers towards the acceptance of the research and the new knowledge claims presented. The interpretation (Move 4-Step 1), contributions (Move 4-Step 4) and limitations of the findings (Move 4-Step 5) are associated with authorial comments conveying relevance (examples 1, 2 and 3):

(1) Perceived usefulness and sense of belonging had *significant* effects on participants' intention to use the system. (Computing)

(2) Achieving this magnitude of strength at such a high homologous temperature of 0.7 Tm is very *significant*. (Nanotechnology)

(3) The results on Ta presented above confirm that microstructural refinement is *significant*, and there is room for improvement. (Nanotechnology)

When evaluating results (Move 4-Step 4), *significant* often accompanies nouns of positive meaning (*savings, advantages, increase, rise)* and negative meaning (*decrease, mismatch, degradation, problems),* thus rendering either a positive or negative judgement and reflecting the writer's favourable or unfavourable position towards the information reported.

Significant is also used in recommendations for further research addressed to the scientific community (Move 7-Step 2, example 4):

(4) This is encouraging, but complacency is not called for. Very *significant* research, engineering, and societal barriers remain before medical robots have widespread impact on health care. (Robotics)

Other common attitudinal adjectives of appreciation in RA discussions are *important* and *interesting*. Although the numbers of occurrences of both adjectives are dissimilar, they are used in statements with the same communicative purpose. Like *significant,* these adjectives are found in a great number of instances in the *Commenting on results* move (Move 4, examples 5 and 6):

(5) This is an *important* result as it provides a reason why {100} B2 type interfaces are observed. (Nanotechnology)

(6) Some *interesting* observations can be made from this figure. (Telecommunications)

These adjectives often collocate with nouns that refer to properties and specific aspects of the objects under study, thus reinforcing the positive subjective view of the author with respect to the interpretation, evaluation and contributions of the results of the study (examples 7, 8, 9, 10 and 11):

(7) Indeed, a *very important property* of the proposed method is that the number of operations increases linearly with the number of elements and obstacles involved in the avoidance checks. (Robotics)

(8) This is an *important* and novel *contribution*. (Robotics)

(9) From these results we conclude that the two most *important aspects* in producing Ni islands are the annealing temperature and the initial Ni layer thickness. (Nanotechnology)

(10) An *interesting feature* of the scale of the microstructure produced by drawing is the resultant increase in solubility. (Nanotechnology)

(11) A more *interesting phenomenon* due to the nonadaptability of the static schemes is that the expected fairness index does not monotonically increase with the number of layers. (Telecommunications)

But they are also common in background statements, especially when referring to the issues analysed in the study (Move 1-Step 5, examples 12 and 13):

(12) After having presented the planning and obstacle avoidance method, some *important* issues regarding its use are addressed. (Robotics)

(13) Experiments previously discussed have yielded some *interesting* insight regarding the reactivity of nanoparticles. (Nanotechnology)

Other uses of these adjectives are found when evaluating the study (Move 6-Step 2, examples 14 and 15):

(14) Since wavelength converters are rather expensive, this study is *important* and necessary. (Telecommunications)

(15) The study of Ni-based layers is therefore *interesting* in its own right since it is at the interface between the strongly and weakly reacting metals. (Nanotechnology)

Finally, they are also used in deductions from the study to recommend further research (Move 7-Step 2, examples 16 and 17):

(16) It will become more and more *important* to design systems that can incorporate a wide range of biomedical sensors and that can work with multiple imaging modalities. (Robotics)

(17) An *interesting* route to extend this research is to assess the impact of these variables on sustainability at a finer level in a more controlled environment. (Computing)

And for drawing implications (Move 7-Step 3, examples 18 and 19):

(18) A key feature of the present method is that it is independent of the quality of tessellation, which is very *important* for practical applications. (Computing)

(19) These results present *interesting* implications. (Computing)

According to the totals in Table 1, the third most common evaluative adjective in the corpus is *effective*. However, this adjective is mainly found in the Nanotechnology sub-corpus (24 instances out of 37), where it is often used when commenting both on techniques and procedures (with *model, approach, detection, visualisation, way,* as in example 20), and on specific properties (with *colour, thickness, rate*, as in example 21):

(20) The fewer the constraints there are, the more *effective* the *approach* is. (Robotics)

(21) White background blue is one of the most *effective colours* for text. (Computing)

It also appears in suggestions for further research (Move 7-Step 2, example 22):

(22) Use of our device in combination with other methods, such as perfusion in the iliac bone, might prove more *effective*. (Robotics)

Critical is another common adjective in Nanotechnology for background information on methodological matters (Move 1-Step 4, example 23), comments on measurements taken into account (Move 4, example 24) and for implications of the study (Move 7-Step 3, example 25):

(23) Moreover, it is *critical* to select the precursor with highly polymerization or strong coagulation ability for forming the shell particle. (Nanotechnology)

(24) [...] the distance from the cluster to the fluorophore is very *critical*, with quenching being not the only problem. (Nanotechnology)

(25) Using an array of such nanoelements may provide better statistics and improve the reliability, which is *critical* for practical applications. (Nanotechnology)

Appreciation is also expressed in our corpus by means of evaluative adverbs, such as *significantly, interestingly, successfully* and *importantly* (see Table 2). The figures clearly indicate that *significantly* (36 instances) is the most frequent in all disciplines.

	Comp	Telecom	Nanotech	Robot	Totals
significantly	3	10	17	6	36
successfully	0	0	2	5	7
interestingly	2	1	2	0	5
importantly	2	0	0	0	2

Table 2. Number of occurrences of evaluative adverbs.

Evaluative adverbs, either pre/postmodifying a verb or used as adjuncts, are employed in our corpus to argue that a finding is positive, interesting or important (Move 4-Step 4, as in examples 26, to 29):

(26) Adaptive-feed rate interpolation *improves* the accuracy *significantly* compared to speed-controlled interpolation. (Computing)

(27) The helicopter *successfully went into* search mode, and when the helipad was visible again, it was able to *successfully track* the helipad, orient appropriately, and land on it. (Robotics)

(28) *Interestingly*, the curves for BPHT in Figs. 7 and 8 indicate that with an appropriate wavelength granularity, BPHT performs the best. (Telecommunications)

(29) More *importantly*, with Likert as first response format, the effect of orientation support was significant, irrespective of response format. (Computing)

Intensifying adverbs

Expressions of appreciation are reinforced and emphasised with intensifying adverbs, such as *strongly* (30 occurrences), *especially* (29 occurrences), *highly* (24 occurrences), *dramatically* (8 occurrences), and *considerably* (7 occurrences), as shown in Table 3.

	Comp	Telecom	Nanotech	Robot	Totals
strongly	3	1	23	3	30
especially	7	5	10	7	29
highly	2	3	11	8	24
dramatically	2	1	2	3	8
considerably	1	0	4	2	7

Table 3. Number of occurrences of intensifying adverbs.

Although the global figures show that *strongly* and *especially* are the most common items in the corpus, the majority of the instances are found in the Nanotechnology sub-corpus and there is variety as to the most frequently used item in each field of knowledge. All these adverbs occur mainly when interpreting results (Move 4-Step 1) but it is worth mentioning the concordances in which they are found. *Strongly* and *dramatically* collocate with verbs such as *emphasise, increase* and *change* to highlight interesting results, as in examples 30, 31 and 32:

(30) The auto-correlation function is much more narrowly focused than the unpreprocessed case and *strongly emphasises* vertical structure. (Robotics)

(31) We can see that with the increasing of the number of faces of the operands, the ratio *increases dramatically*, i.e. the 3D-MSSD algorithm becomes more advantageous. (Computing)

(32) [...] the simple preprocessing that we have implemented *dramatically changes* the correlations to which the model is exposed, altering development. (Robotics)

In contrast, *especially, highly* and *considerably* usually collocate with adjectives either to interpret results (Move 4-Step 1, example 33), to evaluate results (Move 4-Step 4, example 34) or to suggest applications from the research (Move 7- Step 3, example 35):

(33) These are *especially useful* in WANs and arrange receivers into a hierarchical structure of regions. (Computing)

(34) This effect can be seen in Fig. 11, in that the initial error is *considerably larger* in X than in Y and Z. (Robotics)

(35) Alternatively, gossiping can also be used to create a *highly robust* data distribution service. (Telecommunications)

Attitudinal stance adverbs

According to Appraisal theory, attitudinal stance adverbs evaluate claims in affectual terms, which foreground authors' subjective presence. As shown in Table 4, *unfortunately, surprisingly* and *fortunately* are efficient resources that convey feelings or expectations in the corpus. However, the low numbers of occurrences of these adverbs (only 9 instances) leads us to infer that scientists recur to other strategies to represent their attitude within a particular context.

	Comp	Telecom	Nanotech	Robot	Totals
unfortunately	2	1	0	1	4
surprisingly	1	1	0	1	3
fortunately	0	1	0	1	2

Table 4. Number of occurrences of attitudinal stance adverbs.

The writer's attitude to the value of research related to expectations is expressed through attitudinal stance adverbs when comparing results (Move 4-Step 2), as in example 36:

(36) *Surprisingly*, these values are much better than the values obtained in the above landing experiments. (Robotics)

Unfortunately denies the positive expectations set up by the preceding text, thus indicating disagreement with other researchers' claims (example 37):

(37) *Unfortunately*, none of these strategies appear to have been entirely satisfactory. (Computing)

Fortunately is used in the opposite way. In this case, it is negative expectations set up by the preceding text which are denied (example 38):

(38) *Fortunately*, the trigonometric terms in the above nonlinear equations can easily be transformed to polynomials using the well-known tan substitution. (Robotics)

3.2. Expressing certainty

Certainty markers express certainty and conviction as regards the interpretation of results. Koutsantoni's (2004) classification includes in this category certainty adjectives and adverbs, verbs such as *will, going to, demonstrate* or *show*, and discourse-based expressions of confidence, which have the effect of imposing views on readers and contribute to creating a climate of complicity and working towards the acceptance of claims. They serve to establish a relationship with the reader, who is considered to be able to understand and follow the author's reasoning.

Table 5 shows that the most common certainty adjective in the corpus is *clear* (30 instances), followed by *apparent* (25 instances), *evident* (15 instances), *obvious* (15 instances) and *certain* (only 1 instance). However, there are differences across the four disciplinary domains. Nanotechnology concentrates most of the instances. In contrast, the number of occurrences of these adjectives in Telecommunications and Computing is low.

	Comp	Telecom	Nanotech	Robot	Totals
clear	3	2	17	8	30
apparent	2	0	15	3	20
evident	0	1	11	3	15
obvious	2	1	6	6	15
certain	0	0	1	0	1

Table 5. Number of occurrences of certainty adjectives.

Certainty adjectives appear in statements that fulfil different communicative purposes. They are used for background information (Move 1-Step 1, examples 39 and 40):

(39) Several proposals [28, 29] suggest using an active networking approach to provide an AGCS. *It is clear that* if an active networking service is available in each potential node, because of the flexibility it offers, deploying an AGCS becomes an easy task. (Telecommunications)

(40) The most *obvious* approach is binding the analyte (preferably a biocomponent) to the cluster and making the presence or absence of the clusters in the SEA set-up dependent on a biorecognition process. (Nanotechnology)

But they are common in comments on results (Move 4), evaluations of the study (Move 6) and deductions from the research (Move 7). With certainty adjectives authors emphasise their conviction regarding what has been accomplished. When interpreting results (Move 4-Step 1), their use is based on evidence provided by the findings themselves, often displayed in visuals – tables and figures – and examples (see examples 41 to 44):

(41) Also, from the micrograph, it is clear that there are regions where crystallization is not complete. (Nanotechnology)

(42) This fact is also *evident in Tables III-IV*. (Telecommunications)

(43) *From figure 3(a) it is apparent* that there is some evidence of island growth for an initial Ni thickness of 3 nm with a mean diameter of 67 nm. (Nanotechnology)

(44) By active sensing the robot is approaching the beacons (*obvious from the results in the examples*), thereby deviating from a reference trajectory. (Robotics)

Certainty adjectives are also used when comparing results (Move 4-Step 2, example 45):

(45) This is *evident after comparing the results* for the 200 and 485 nm thick specimens. (Nanotechnology)

By means of a direct and committed language, the author stresses the achievements of the research carried out so as to consolidate her/his research space. The use of negative statements, however, conveys honesty and modesty on the part of the author (examples 46 and 47):

(46) *It is clear* from our studies, and those of others, that rapid surface restructuring can occur on gold surfaces. (Nanotechnology)

(47) *It is not clear* why increasing the temperature of bonding from 1090 to 1185°C failed to give a significant increase in the TD of the bond interlayer. (Nanotechnology)

As regards certainty adverbs, Table 6 shows that *clearly* is the most commonly used item (39 occurrences).

	Comp	Telecom	Nanotech	Robot	Totals
clearly	3	4	23	9	39
certainly	3	0	1	7	11
undoubtedly	0	3	0	1	4

Table 6. Number of occurrences of certainty adverbs.

Certainty adverbs express certainty and conviction for the interpretation of results (Move 4-Step 1, example 48) and the comparison with other authors' claims (Move 4-Step 2, example 49):

(48) The above results *clearly* show hierarchical attention selectivity and appropriated believable performance in a complicated natural scene. (Robotics)

(49) This is *clearly* an advantage of the method against heavy cell-decomposition approaches where the computational burden is high. (Robotics)

Such formulations either present propositions as given, that is, as generally known or accepted (example 50), or emphasise specific qualities of the objects under study, as in example 51:

(50) A per-source shortest path tree is *undoubtedly* more efficient than a single shared tree used by many different sources. (Telecommunications)

(51) The age-based message purging performance of the pbcast algorithm is *clearly* higher. (Nanotechnology)

3.3. Epistemic expressions

Epistemic expressions are also used to convey the author's positioning on the facts stated. Epistemic lexis can be included in the category of hedges, which are useful ways of softening assessments and preventing counter-claims (Hyland 1998b). Together with certainty expressions, epistemic devices pertain to the engagement appraisal subsystem. Moreover, they attenuate the evaluative/interpretative compromise and contribute to presenting hypotheses with caution when evidential knowledge places limitations on factual reporting (Soler 2002).

	Comp	Telecom	Nanotech	Robot	Totals
likely	13	1	20	1	25
possible	0	0	5	9	14
unlikely	1	0	0	2	3
probable	0	0	2	0	2

Table 7. Number of occurrences of epistemic adjectives.

Table 7 shows the number of epistemic adjectives in the corpus. *Likely* is the most common epistemic adjective (25 instances), followed by *possible* (14 instances). *Unlikely* and *probable*, however, are scarcely used. Comparison of Tables 5 and 7 allows us to infer that authors in Nanotechnology combine certainty expressions with less categorical lexis. This combination in a text reflects a strategy of balancing statements of conviction that highlight the achievements of the research with speculative personal statements seeking acceptance by the scientific community and showing respect to viewpoints different from the author's. However, the figures obtained in Computing seem to indicate that authors in this discipline recur to speculative devices rather than certainty expressions when referring to their findings, which renders discourse unassertive. The low number of certainty and epistemic adjectives in Telecommunications suggests that authors in this area make use of other markers.

Epistemic adjectives are mitigating resources for projecting caution when interpreting (Move 4-Step 1) and accounting for results (Move 4-Step 3), in which case *likely* collocates with *cause, source, due to, owing to*, while *possible* and *probable* collocate with *explanation, effect* and *difference.* A selection of these collocations is presented in examples 52, 53 and 54.

(52) They often express frustration with minor annoyances in the environment and trade tips with each other about problems they have recently solved. *There are two likely causes of this*. (Computing)

(53) Thus, the greater stability of EGBDs at CSLBs appears to be due to a decreased vacancy diffusivity in those boundaries, *most likely owing to* their more ordered geometric structure. (Nanotechnology)

(54) In general, Tg is lower in the reinforced material compared to the unreinforced material (Fig. 6). One *possible explanation* for this difference is that the kinetics of the glass transition are altered by the presence of the particles. (Nanotechnology)

Epistemic adjectives are also typically used for deductions from the research (Move 7), both to suggest further research (Move 7-Step 2, example 55) and to draw implications from the study (Move 7-Step 3, example 56):

(55) [...] it is *likely* that glyph performance will improve beyond the levels suggested by our results. Performance is also *likely* to improve in an applied setting, like stock markets and military environments. (Computing)

(56) *Possible* fields of application include food-component identifications, allergen screening or the detection of pathogenic bacteria. (Nanotechnology)

Other expressions that mark the writer's confidence in the truth of a proposition are epistemic adverbs. They acknowledge subjective uncertainties, and are motivated by the writer's desire to explicitly convey an assessment of the reliability of propositional validity. As Table 8 shows, *probably* is the most common epistemic adverb (27 occurrences), followed by *possibly* (18 occurrences) and *presumably* (8 occurrences). As with epistemic adjectives, Nanotechnology concentrates the greatest number of instances.

	Comp	Telecom	Nanotech	Robot	Totals
probably	9	1	13	4	27
possibly	0	1	17	0	18
presumably	0	0	8	0	8

Table 8. Number of instances of epistemic adverbs.

The examples in the corpus are concerned with likelihood when interpreting (Move 4-Steps 1, examples 57, 58 and 59) and accounting for results (Move 4-Step 3, examples 60, 61 and 62):

(57) However, the general conclusion from our experiments is that the Bethe approximation *probably* breaks down before any significant difference between the two methods shows up. (Robotics)

(58) The primary creep as demonstrated in Figs. 1 and 2 is *possibly* corresponding to the transient period for the establishment of steady state condition of diffusional fluxes (Nanotechnology)

(59) [...] the mesoscale-dimples would *presumably* be associated with both concentrating effect of smaller grains and increasing deformation ability in larger grains. (Nanotechnology)

(60) The flexural strength is affected by the carbon contents (or the SiC contents), low addition of carbon black results in a considerable high strength. This is *probably* attributed to the formation of a high fraction of fibrous Si3N4 grains microstructure in sample with low carbon content (also low SiC content). (Nanotechnology)

(61) During the unloading only one discontinuity or a short step was observed [...], *possibly* due to the residual deform of thin film. (Nanotechnology)

(62) *Presumably* this is why Al and Nb segregate more strongly to these defects. (Nanotechnology)

4. Conclusion

The present study has looked at appraisal resources used by engineering authors in an attempt to establish a relationship between lexical choices and communicative purposes in a small corpus of Computing, Telecommunications, Nanotechnology, and Robotics RA discussion sections. The items expressing authors' attitude and engagement towards propositions have been selected and two grammatical categories have been investigated: adjectives and adverbs.

Nanotechnology RAs show the highest use of appraisal resources in the corpus, while the figures in the Telecommunications RAs indicate that such devices are not frequently employed in this discipline. Attitude markers form the most common group of resources in the corpus, whereas the number of epistemic adjectives and adverbs is the lowest. This suggests that writers recur to other forms of hedging, probably modals. Evaluative lexis is typically associated with authorial comment, for interpreting and evaluating results, and concentrates the highest number of occurrences. Adjectives clearly predominate over adverbs: *significant* and *important* are the most common evaluative expressions in the corpus, and are used to describe the relationships among observed phenomena and the significance of the collected results. Certainty expressions also contribute to enhancing the authors' merits: they are used for interpreting, accounting for and comparing results, and guide the reader through the reasoning. The adjectives *clear, apparent, obvious* and *evident* and the adverbs *clearly* and *certainly* most often serve the writers' purpose to create a text where their claims seem unproblematically related to observed data. In contrast, the mitigating expressions *likely, probably, possibly* and *possible* are the most usual means of projecting modesty and caution. Although they are found in suggestions for further research and applications, they are typically used as hedges marking the writers' reluctance to present or explain propositional information categorically and showing respect for colleagues' different views.

This exploratory study has hopefully provided valuable information about the choices of attitude, certainty and epistemic adjectives and adverbs and their implications in scientific discourse. Future research on discourse and grammatical resources is needed to complete the appraisal network of the most common resources in RAs from the fields of Computing, Telecommunications, Nanotechnology and Robotics. We think these findings may be of great use both for the teaching of EAP and for awareness about written language in specific contexts.

References

Chafe, Wallace L. / Nichols, Johanna (eds) 1986. *Evidentiality: The Linguistic Coding of Epistemology.* Norwood, NJ: Ablex.

Conrad, Susan / Biber, Douglas 2003. Adverbial Marking of Stance in Speech and Writing. In Hunston, Susan / Thompson, Geoffrey (eds) *Evaluation in Text. Authorial Stance and the Construction of Discourse.* New York: Oxford University Press, 56-73.

Dudley-Evans, Tony 1994. Genre Analysis: An Approach to Text Analysis for ESP. In Coulthard, Malcolm (ed.) *Advances in Written Text Analysis.* London: Routledge, 219-228.

Gil-Salom, Luz / Soler-Monreal, Carmen / Stuart, Keith 2001. Lexico-grammatical Resources of Evaluation in Research Articles. *Trabajos en Lingüística Aplicada.* Barcelona: Univerbook S.L., 329-337.

Gil-Salom, Luz / Soler-Monreal, Carmen / Stuart, Keith / Candela-Coderch, Jorge 2004. *Textworks* (CD-Rom). Valencia: Universidad Politécnica de Valencia.

Halliday, Michael Alexander K. 1994 (1985). *An Introduction to Functional Grammar.* London: Edward Arnold.

Hunston, Susan 1994. Evaluation and Organisation in a Sample of Written Academic Discourse. In Coulthard, Malcolm (ed.) *Advances in Written Text Analysis.* London: Routledge, 191-218.

Hunston, Susan 2003 (1999). Evaluation and the Planes of Discourse: Status and Value in Persuasive Texts. In Hunston, Susan / Thompson, Geoffrey (eds) *Evaluation in Text. Authorial Stance and the Construction of Discourse*. New York: Oxford University Press, 176-207.

Hyland, Ken 1998a. Persuasion and Context: The Pragmatics of Academic Metadiscourse. *Journal of Pragmatics*. 30, 437-455.

Hyland, Ken 1998b. Boosting, Hedging and the Negotiation of Academic Knowledge. *Text*. 18/3, 349-382.

Hyland, Ken 1998c. *Hedging in Scientific Research Articles*. Amsterdam: John Benjamins.

Hyland, Ken 2001. Bringing in the Reader: Addressee Features in Academic Articles. *Written Communication*. 18/4, 549-574.

Hyland, Ken 2002. Activity and Evaluation: Reporting Practices in Academic Practices. In Flowerdew, John (ed.) *Academic Discourse*. London: Pearson Education Ltd., 115-130

Koutsantoni, Dimitra 2004. Attitude, Certainty and Allusions to Common Knowledge in Scientific Research Articles. *Journal of English for Academic Purposes*. 3, 142-175.

Latour, Bruno / Woolgar, Steve 1979. *Laboratory Life: The Construction of Scientific Facts*. Princeton: Princeton University Press.

Lewin, Beverly A. / Fine, Jonathan / Young, Lynne 2001. *Expository Discourse*. London: Continuum.

Martin, James R. 2003. Beyond Exchange: Appraisal Systems in English. In Hunston, Susan / G. Thompson (eds) *Evaluation in Text. Authorial Stance and the Construction of Discourse*. New York: Oxford University Press, 142-175.

Peacock, Matthew 2002. Communicative Moves in the Discussion Section of Research Articles. *System*. 30, 479-497.

Posteguillo, Santiago 1999. The Scientific Structure of Computer Science Research Articles. *English for Specific Purposes*. 18/2, 139-160.

Ruiying, Yang / Allison, Desmond 2003. Research Articles in Applied Linguistics: Moving from Results to Conclusions. *English for Specific Purposes*. 22/4, 365-385.

Ruiying, Yang / Allison, Desmond 2004. Research Articles in Applied Linguistics: Structures from a Functional Perspective. *English for Specific Purposes*. 23/3, 264-279.

Salager-Meyer, Françoise 1994. Hedges and Textual Communicative Function in Medical English Written Discourse. *English for Specific Purposes*. 13/2, 149-170.

Samraj, Betty 2002. Introductions in Research Articles: Variations across Disciplines. *English for Specific Purposes*. 21/1, 1-17.

Skelton, John 1997. The Representation of Truth in Academic Medical Writing. *Applied Linguistics*. 18/2, 121-140.

Soler, Viviana 2002. Analysing Adjectives in Scientific Discourse: An Exploratory Study with Educational Applications for Spanish Speakers at Advanced University Level. *English for Specific Purposes*. 21, 145-165.

Soler-Monreal, Carmen / Gil-Salom, Luz Forthcoming. Moves, Steps and Linguistic Signals in RA Discussion Sections.

Soler-Monreal, Carmen / Gil-Salom, Luz / Carbonell-Olivares, María 2006a. Unconventional Section Headings in Computing and Robotics Research Articles. *Actas V Congreso Internacional AELFE*. Zaragoza: Prensas Universitarias, 77-86.

Soler-Monreal, Carmen / Gil-Salom, Luz / Carbonell-Olivares, María 2006b. A Study of Section Headings in Computing, Robotics and Telecommunications Research Articles. *RaeL*. 5, 1-26.

Stotesbury, Hilda 2003. Evaluation in Research Article Abstracts in the Narrative and Hard Sciences. *Journal of English for Academic Purposes*. 2, 327-341.

Swales, John M. 1990. *Genre Analysis: English in Academic and Research Settings*. Cambridge: Cambridge University Press.

Thompson, Geoffrey / Ye, YiyunYe 1991. Evaluation in the Reporting Verbs Used in Academic Papers. *Applied Linguistics*. 12/4, 365-382. <http://www.grammatics.com/appraisal.html>.

Appendix

Discussion section: Moves/Steps (based on Ruiying and Allison's (2003) two-level macrostructure account for RA discussions)	Discipline and number of discussion sections with the move/step				
	Comp 11	Nanotech 10	Robot 15	Telecom 10	Totals 46
Direct reference to the findings					
Move 1. Background information	11	4	11	8	33
Step 1. Reviewing items of previous research (figures must go a level lower)	3	-	4	-	7
Step 2. Indicating a gap	1	-	2	-	3
Step 3. Restating purposes	3	1	1	5	10
Step 4. Restating method	4	1	4	3	12
Step 5. Specifying issues analysed	2	2	3	7	14
Step 6. Announcing principal findings	2	-	1	-	3
Move 2. Reporting results	7	6	10	10	33
Move 3. Summarising results	2	1	7	2	12
Move 4. Commenting on results	11	10	13	10	44
Step 1. Interpreting results	7	8	10	10	35
Step 2. Comparing results	5	5	8	4	22
Step 3. Accounting for results	7	8	3	9	27
Step 4. Evaluating results	4	1	8	6	19
Step 5. Indicating limitations of results	-	1	2	1	4
Step 6. Exemplifying	1	-	1	1	3
More general statements on the study					
Move 5. Summarising the study	1	1	5	2	9
Move 6. Evaluating the study	3	2	4	3	12
Step 1. Indicating limitations	2	-	3	3	8
Step 2. Indicating significance/advantages	4	-	3	1	8
Step 3. Evaluating methodology	-	2	-	1	3
Move 7. Deductions from the research	11	2	10	3	26
Step 1. Planning further research	2	1	3	2	8
Step 2. Recommending further research	5	1	5	1	12
Step 3. Drawing implications	1	-	6	-	7

Table 9. Presence of moves and steps in RA discussion sections (Soler-Monreal/Gil-Salom forthcoming).

Concepción Orna-Montesinos

Hyponymy Relations in Construction Textbooks: A Corpus-Based Analysis

1. Introduction

The starting point for the analysis of disciplinary lexis is the assumption that different discourses structure the world in a different way, and that the members of a particular speech community share a common understanding of disciplinary vocabulary. Vocabulary choice is thus subject to situation and context constraints (Carter/McCarthy 1988; Fairclough 2003; Bloomer *et al.* 2005).

According to semantic theory, lexical relations are specific to a particular discourse community: the words may be the same but with a different use (Fairclough 2003) and it is through the presupposed structuring of disciplinary lexical knowledge that we can make sense of texts. Along similar lines, other authors such as Saeed (1997: 63) argue that the "organizational principle in the lexicon is the lexical field, [...] a group of lexemes which belong to a particular activity or area of specialist knowledge".

As different analyses of scientific and technical discourse have shown (Trimble 1985; Nation/Kyongho 1995; Coxhead/Nation 2001; Chung/Nation 2003), one of the main problems this discourse poses concerns its lexis, particularly technical vocabulary – "specific to a particular topic, field or discipline" (Coxhead/Nation 2001: 261) – and sub-technical vocabulary – "words that have the same meaning in several scientific or technical disciplines" and "those 'common' words that occur with special meanings in specific scientific and technical fields" (Trimble 1985: 129). From a pedagogical perspective (Trimble 1985; Carter/McCarthy 1988; Thurstun/Candlin 1998; Mudraya 2006; Chen/Ge 2007), subject-specific vocabulary has been considered key for disciplinary-knowledge development, since specific vocabulary is associated with particular genres. It is also important to raise learner's awareness that specialised vocabulary acquisition is not restricted to learning subject-specific terminology, but also to being able to interpret and to use specialised/discipline-specific academic vocabulary. As Carter/McCarthy (1988) suggest, vocabulary choice is part of the sociolinguistic or pragmatic competence of the learner, who should develop some lexical awareness of the disciplinary constraints, i.e. participants, goals, settings, and of its implied interpersonal factors, i.e. formality or social distance.

Drawing on these premises, the purpose of this paper is to analyse the semantic relations which contribute to organising the specialised discourse of architecture, construction and civil engineering, and to determine how semantic relations condition the choices available to the members of this discourse community to lexicalise their disciplinary world. Based on a 1,000,000-word corpus of textbook samples (the *Construction Textbooks Corpus,* hereafter CTC) contrasted with the WordNet project database, I seek to define what a *building* is and to analyse the semantic relations of hyponymy, i.e. the types of building, which are relevant to the discipline. In addition, I attempt to determine how hyponyms offer the members of this speech community the possibility of choosing between more general or more specific meanings, and how these meanings can be adapted to the specialised register of disciplinary discourse.

2. Semantic relations

Semantics has been traditionally defined as the study of the meaning of words in a language (Kempson 1977; Lyons 1977; Cruse 1986; Hudson 1995; Salkie 1995; Saeed 1997; Kreidler 1998; Widdowson 1998; Gregory 1999; Jackson/Amvela 2000; Yule 2001; Bloomer *et al.* 2005; Kearns 2006). However, words do not appear in isolation, but rather in the more complex units and grammatical relations of a linguistic system. As Cruse (1986: 1) puts it: "words are not the bearers of messages; they do not, of themselves, 'make sense'". As a "container of meaning" (Yule 2001: 118), the word contributes to the meaning of phrases and sentences which, in turn, determine their semantic properties through contextual relations. Therefore, the distinction between the multiple semantic uses of a word depends on lexical relations.

For Cruse (1986: 16), lexical relations reflect "a pattern of affinities and disaffinities with all the other words in the language". These affinities are commonly classified into paradigmatic and syntagmatic relations (Lyons 1968, 1977; Cruse 1986, 2000, 2003; Jackson/Amvela 2000; Fairclough 2003). Paradigmatic relations are "systems of choices a speaker faces when encoding his message" (Cruse 1986: 86). These relations "in absentia" (Jackson/Amvela 2000: 12) explain how words can be interchanged in different linguistic positions, since the use of a word in a context suggests other substitutable or alternative words; for example, the noun *building* is paradigmatically associated with such nouns as *cathedral, erection*, *room* or *masonry*. Syntagmatic relations, on the other hand, are "discourse cohesion devices which add necessary informational redundancy to the message" (Cruse 1986: 86). These are relations "in praesentia" (Jackson/Amvela 2000: 13), which involve their use as signalling, disambiguation or metalinguistic devices. For instance, there is a syntagmatic relationship – usually grammatical – between the noun *skyscraper* and the verbs *construct, build* or *demolish* and with adjectives like *high* or *modern* in utterances such as *A skyscraper was constructed* or *It's the highest skyscraper in the world.*

As Cruse (2003: 259) further states: "the vocabulary of a language is not just a collection of words scattered randomly through semantic space: it is at least partially structured by recurrent sense relations". Sense is defined as "the set, or network, of sense-relations that hold between it and other expressions of the same language" (Lyons 1995: 80). A word sense is the basic unit of meaning, and sense relations are relations between word meanings. Words acquire their meaning in the network of semantic relations established with other words in the language. Jackson/Amvela (2000) borrow Saussure's model to represent the network of associations of form and meaning between words. Graph 1 adapts their diagram to the associations of the noun *building*: in (1) *building* is connected with the verb forms *build* and *built* both semantically and formally; the nouns *construction* and *house* are linked in (2) on account of their semantic similarity; the common ending *–ing*, which turns the verbs *build, mean* and *begin* into nouns, associates *building* with *meaning* and *beginning* in line (3); finally, line (4) connects *building* with *bring* and *interesting* because of the coincidental common ending.

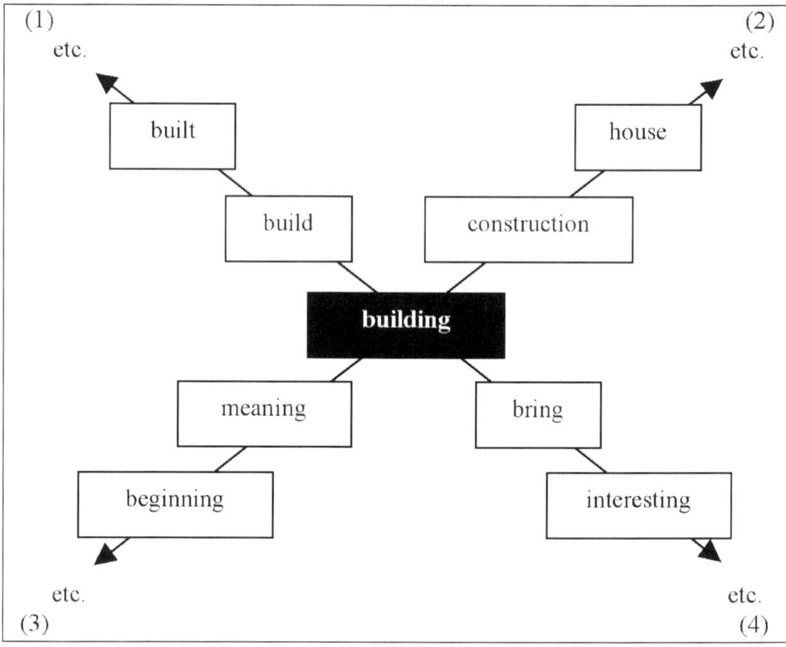

Graph 1. Word associations of *building*. Adapted from Jackson/Amvela (2000: 12).

Cruse (1986: 87) identifies four basic sense relations: identity (class A and class B have the same members); inclusion (class B is wholly included in class A); overlap (class A and class B have members in common but each has members not found in the other); and disjunction (class A and class B have no members in common). Identity and inclusion relations encompass hyponymy (*building* → *skyscraper*), meronymy (*building* → *corner*) and synonymy (*building* → *edifice*), whereas incompatibility (*hotel* → *hospital*), co-meronymy (*roof* → *floor*) and opposites (*erect* → *demolish*) are relations of opposition and exclusion.

3. Hyponymy relations

Given its role in structuring and articulating semantic space as well as in organising the lexicon, hyponymy is generally considered as the most important paradigmatic relation (Lyons 1968, 1977; Cruse 1986, 2000, 2006; Saeed 1997; Widdowson 1998; Yule 2001; Darian 2003; Bloomer *et al.* 2005; Griffiths 2006).

According to Lyons (1977: 292), hyponymy is the 'X is a type of / kind of / sort of' relation. If we apply this definition to the language of construction, *house* is a 'hyponym' of the 'hypernym' *building* (also called 'hyperonym' or 'superordinate') if it is logically acceptable that *A house is a kind of building*. Hyponymy is a class-inclusion, or encapsulation, relation because the class of *houses* is included in the class of *buildings*. The 'hyponym' (included category) *house* includes the category of the 'hypernym' (inclusive category) *building* so that if something is a *house*, it is necessarily a *building,* and the meaning of *house* includes the meaning of *building*. Following Miller (1998), the sentence *They constructed a bungalow* does not need to specify that *They constructed a building*, or that it has a roof and walls, or that it is a dwelling, because *bungalow* has inherited the properties of *house*, which, in turn, has inherited the properties of *building*.

Hyponymy establishes a unidirectional hierarchy – in contrast with the bidirectional one in the case of synonymy – as a hyponym can be the hypernym of another, so that a *house* is a hyponym of *building* and a hypernym of *bungalow*, the next level of classification. The tree can be extended to subsequent levels in a relation of transitivity, until no distinctive features remain (Lyons 1977; Cruse 1986): if *bungalow* is a hyponym of *house*, which in turn is a hyponym of *building*, it follows that a *bungalow* is a hyponym of *building*. In other words, the meaning of the hyponym inherits the meaning of the hypernym and adds certain specificity: *bungalow* inherits all the properties of its hypernym *house* and *house* inherits the properties of the hypernym *building*, but each one adds its own specification (see graph 2).

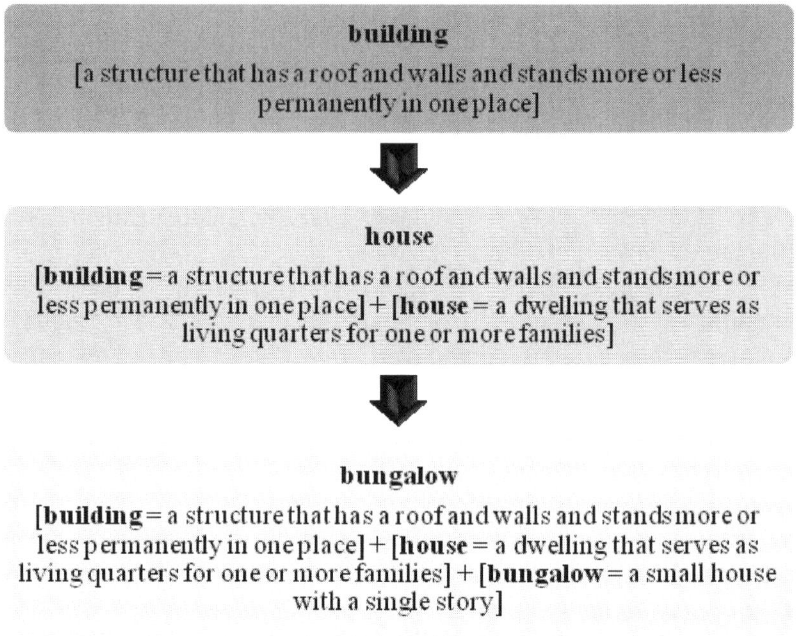

Graph 2. Inheritance of meaning.

As can be observed, hyponymy establishes a relation of sense which is explained by Lyons (1968) in terms of hierarchical implication. The semantic relations of a word translate into the entailments that the word will suggest when used in a sentence. A sentence containing a hyponym unilaterally entails an identical sentence in which a superordinate is used: *They constructed a library* implies that *They constructed a building*, although *They constructed a building* does not imply that *They constructed a library*.

Hudson (1995: 16) refers to what might seem a "grossly extravagant" range of hyponyms and asks "[w]hy should a language provide so many overlapping classifications, when just one of them would do?". It is the aim of this paper to analyse the role of a myriad of hyponyms in the discipline of construction: why a *building* can be referred to in *n* different ways and why the same entity, a *building,* can be referred to as a *club* or a *school.* To answer this question, we must refer back to the definition of hyponyms in terms of increasing specificity: specific discourse requires specific vocabulary. Drawing on Jackson/Amvela (2000), a *school* is a more specific designation of *building*: *building* can be used to designate constructions which are not *schools* but which share the essential features of 'buildingness' ('a structure that has a roof and walls and stands more or less permanently in one place').

Cruse (1986: 153) argues that "[o]ne of the linguistic variables under our control as speakers of a language is the semantic specificity of expressions", and therefore, we can choose to use a word with a light semantic load, as in *They constructed a building,* or we can choose to "pack in a lot of meaning" (Cruse 1986: 153), as in *They constructed a cathedral.* A hyponym is generally loaded with meaning and carries particular connotations (i.e. *Pentagon,* which restrict its use to a specific area of meaning (Carter *et al*. 2001). The use of the hypernym *building* simply refers to the event of construction. The specification conveyed by the hyponym *cathedral*, however, not only makes it explicit which type of building, but adds the entailment of 'religious use', and further potential meanings of 'architectural style', 'size', 'age', etc. *They constructed a cathedral* is certainly "more informative" than *They constructed a building* (Kreidler 1998: 93).

Co-hyponyms have different marked and unmarked meanings for style or field. Marked co-hyponyms have "more meaning" and "it is this extra meaning which restricts their collocational normality" (Cruse 2002: 8). In *They constructed a barn* and *They constructed a courthouse* the meaning of the co-hyponym *barn* implies semantic associations with a relatively small, not sophisticated type of building meant for private use and usually located in a rural area; on the other hand, the co-hyponym *courthouse* suggests a usually big and complex type of construction, an example of civil construction oriented to public use in larger urban areas. We can expect *barn* to be found in texts dealing with agricultural or farming construction and collocating with such words as *farm*, *cattle*, *grain* or *store,* whereas *courthouse* would appear in civil or historical construction texts in collocation with, for example, *justice*, *government* or *trial.* As Cruse (1986: 153) puts it:

> To communicate effectively, we must submit to certain constraints on the choice of level of specificity in particular situations [...] The amount of information (and hence, in general, the degree of specificity) that is required will vary from situation to situation.

Hyponym/superordinate chains offer the possibility of choosing between more general or more specific meanings, which can be adapted to the specialised register of disciplinary discourse (Lyons 1974: 455). The members of a speech community share "encyclopaedic knowledge" (Jackson/Amvela 2000: 103) about words, which will allow them to adapt to the necessary situation of specificity in disciplinary contexts, in which a *temple*, for instance, must be named a *ziggurat* ('a rectangular tiered temple or terraced mound erected by the ancient Assyrians and Babylonians'), or a *cathedral* must be precisely designated as a *minster* ('any of certain cathedrals and large churches; originally connected to a monastery'). Similarly, *The shed was demolished* does not entail the same consequences as *The Houses of Parliament were demolished.* For a construction engineer or architect, it definitely matters whether he/she receives *a commission for building a cottage* or *a commission for building a mansion,* and an architect should weigh totally different considerations for *the design of a brasserie* or for *the design of a resort.*

4. A word net of *building*

This section offers the preliminary results of the analysis of the semantic relations of the noun *building* in the CTC. To do so, the frequency analysis of the hyponyms and meronyms of the noun *building* in the corpus was contrasted with the set of cognitive synonyms, hyponyms and meronyms provided by the lexical database of the WordNet project (Fellbaum 1998). The WordNet project (<http://wordnet.princeton.edu>), developed at Princeton University under the direction of George Miller, is a large lexical database of English, in which words are grouped into sets of cognitive synonyms (synsets) interlinked by means of conceptual-semantic and lexical relations. Unlike the traditional alphabetical organisation of dictionaries in terms of form, WordNet, inspired by psycholinguistics, organises the lexicon in terms of meaning and can thus be searched conceptually. Adopting a list of 25 semantic primes (beginners of a hierarchy[1]), WordNet maps a complex network of semantic hierarchies linked by semantic relations of synonymy, antonymy, hyponymy and meronymy, and organised in hyponym/hypernym and meronym/holonym trails.

A search of the noun *building* in WordNet yields the following definitions:

1. *building*, *edifice* (a structure that has a roof and walls and stands more or less permanently in one place)
2. *construction*, *building* (the act of constructing something)
3. *construction*, *building* (the commercial activity involved in repairing old structures or constructing new ones)
4. *building* (the occupants of a building)

Four different senses of the word are distinguished and, consequently, four synonym sets are established. The use of a synonym ({*building, edifice*}, {*building, construction*}) serves to resolve polysemy, and clearly differentiates the intended meaning of sense 1 against sense 2 and sense 3. However, a short definition is provided to disambiguate senses 2 and 3, where the same synonym is provided, or sense 4, a single member synset with no synonyms available.

A previous study of a sample of the corpus (Orna-Montesinos 2008) showed the prevalence of sense 1 (in 58.70% of the cases), in contrast with sense 3 (27.50% of the occurrences) and sense 2 (in 12.60% of the analysed samples); there were no occurrences of sense 4, 'the occupants of a building', although a different meaning was found in 1.21% of instances, which mean 'the art of fabricating edifices'.

1 {act, action, activity} – {animal, fauna} – {artifact} – {attribute, property} – {body, corpus} – {cognition, knowledge} – {communication} – {event, happening} – {feeling, emotion} – {food} – {group, collection}- {location, place} – {motive} – {natural object} – {natural phenomenon} – {person, human being} – {plant, flora} – {possession} – {process} – {quantity, amount} – {relation} – {shape} – {state, condition} – {substance} – {time}

Given the importance of the noun *building* as an edifice, I will focus on the analysis of sense 1 of *building*. Buildings are of key importance for professionals in the construction discipline: they are the main economic reason of their profession, but also the main concern of their thinking. Buildings are not only what they do, but also what they write about.

WordNet lists 485 hyponyms of {*building, edifice*} distributed in five levels of hyponymy; 132 of those hyponyms (27.22%) are employed by authors in the CTC, as table 1 shows:

	WordNet	CTC	%
1st hyponymy level	110	42	38.18
2nd hyponymy level	237	66	27.85
3rd hyponymy level	119	17	14.29
4th hyponymy level	13	6	46.15
5th hyponymy level	6	1	16.67
Total	485	132	27.22

Table 1. WordNet/CTC hyponyms of *building*.

As table 2 illustrates, the 132 hyponyms are used 1,781 times. First level hyponyms, more general in character, are considerably more frequent than those in the lower levels, more specific categories:

	Total different hyponyms	Total frequency
1st hyponymy level	42	1,047
2nd hyponymy level	66	569
3rd hyponymy level	17	142
4th hyponymy level	6	18
5th hyponymy level	1	5
	132	1,781

Table 2. Frequency of hyponyms.

Table 3 shows the frequency of occurrence of the most common hyponyms of building.[2]

house (a)*	419	ruin*	19
church**	133	skyscraper*	18
hotel*	126	palace***	17
center/centre*	124	café**	16
hall (b)*	68	cathedral (b)***	15
Theatre/theatre*	66	conference center**	14
temple (c)**	62	library*	14
school*	52	apartment building*	13
hospital**	37	castle***	13
temple (b)**	35	house (b)*	12
greenhouse*	35	pumping station**	12
cathedral (a)***	26	chapel**	11
opera house**	23	terraced house***	11
solar house**	23	restaurant*	10
garage**	21	shed**	10
Parthenon***	21	sports arena**	10
office building*	20		

Table 3. Most common hyponyms of *building.* [3]

House (with 431 occurrences in the first level of hyponymy) is the most common first level hyponym of *building.* The prevalence of *house* in the corpus – 48.99% of first level hyponyms – can be explained using the notion of prototypes. Saeed (1997: 37), citing Rosch (1976), defines prototypes as the typical members of a category, as opposed to the less typical or peripheral members. For Yule (2001: 106), a prototype is "the characteristic instance". Most would accept *house* as the prototype of *building*, certainly more prototypical of the category 'building' than, for example, a *shed* or a *funeral parlor*. Thus, *house* would be at the basic level of the cognitive hierarchy which "is most used in everyday life;

2 For the complete word tree of *building* in the CTC refer to Appendix.
3 *= first level of hyponymy; **= second level;***= third level.

it is acquired first by children; in experiments it is the level at which adults spontaneously name objects, such objects are recognised more quickly in tests, and so on" (Saeed 1997: 39). *House* also has the widest synset in the CTC: with 97 hyponyms in the second level (58, 12 and 5 in the subsequent levels), it is the only hyponym which reaches the fifth level of hyponymy (see table 4).

house (a) (419)	bungalow (2)			
	cottage (7)			
	cabin *	log cabin (1)		
	country house (2)	summer house (1)		
		villa (8)		
		villa (1)		
	detached house (7)			
	farmhouse (2)			
	guesthouse (4)			
	lodge (1)			
	maisonette (4)			
	mansion (4)	palace (17)		
		castle (13)		
		stately home (3)		
		palace (1)	Alhambra (2)	
			Versailles (4)	
	residence (3)	religious residence*	convent (2)	
			monastery (3)	abbey (5)
		Vatican (2)		
	town house (2)	terraced house (11)		
	solar house (23)			
house (b) (12)	bathhouse (3)			
	courthouse (6)			
	sports arena (10)			
	icehouse (5)			
	pumping station (12)			

Table 4. The word tree of *house*.

The prototypical *house* thus has many facets that only an expert with knowledge of the domain can structure hierarchically since, as Saeed (1997: 40) concludes, experts "might have a different, richer structure at subordination levels for these categories from the average person".

The historical predominance of religious construction is reflected in the corpus and, although the hyponym *place of worship* is not frequent, its synset is the second most frequently used in the corpus, with 256 occurrences, 84 and 6 in the lower levels. *Church* is used 133 times and *church building* 5 times. Several other co-hyponyms are common: *temple* (97), *cathedral* (41), *chapel* (11), *basilica* (9), *mosque* (7) and *abbey* (6). Also frequent are commercial construction examples such as *hotel* (126) and its different hyponyms: *hostel* (2), *inn* (2), *lodge* (2), *spa* (1), *resort* (6)) and *office building / office block* (24); examples of institutional construction like *school* (52), *hospital* (37) or *opera house / opera* (30); and polyvalent buildings such as *center/centre* (124) and its hyponyms: *conference center* (14), *community center* (3), *call center* (1), *research center* (1) or *hall* (73).

Some famous historical buildings also appear in the corpus: *Parthenon* (21), *Pentagon* (9), *Versailles* (4), *Chartres Cathedral* (4), *Alhambra* (2), *Vatican* (2), *Sistine Chapel* (2) and *Houses of Parliament* (1). These are all architectural prestigious buildings, frequently considered part of the cultural and historical heritage of the western world and an artistic model for future generations.

Another significant feature of the CTC is the higher frequency of some hyponyms over their hypernym. For instance, although the hypernym *tavern* does not appear in the corpus, its hyponyms *public house* (1) and *pub* (6) do; *outbuilding* is used only once but *garage* (21) and *shed* (10) are significantly more frequent. The higher frequency of these hyponyms indicates that the hypernym is a noun phrase which tries to encompass the meaning of different hyponyms. The corpus shows that textbook authors seem to have a preference for the more specific words rather than for the more general noun phrase. The hyponym *terraced house* is used 11 times, whereas neither of its two synonymous hypernyms *row house* and *town house* is found in the corpus. The same behaviour is shown in the hypernym/hyponym chains: *gambling house* (0) → *casino* (1); *religious residence* (0) → *convent* (2), *monastery* (3); *government building* (2) → *capitol* (2), *courthouse* (5), *embassy*

(4), *Pentagon* (9); *country house* (2) → *villa* (8); *medical building* (3), *health facility* (2) → *clinic* (4), *hospital* (38).

Only 132 of the 485 hyponyms compiled by WordNet (27.22%) are used in the CTC and, despite some notable exceptions, the corpus displays a very low frequency of use of the 132 hyponyms: 56.82% of them are used less than five times throughout the corpus (see table 5), and 24.24%, only once (see table 6).

	Less than 5	Total	%
1st hyponymy level	25	42	59.52
2nd hyponymy level	35	66	53.03
3rd hyponymy level	8	17	47.06
4th hyponymy level	6	6	100
5th hyponymy level	1	1	100
Total	75	132	56.82

Table 5. Hyponyms used less than five times.

	Only once	Total	%
1st hyponymy level	10	42	23.81
2nd hyponymy level	18	66	27.27
3rd hyponymy level	4	17	23.53
4th hyponymy level	0	6	0
5th hyponymy level	0	1	0
Total	32	132	24.24

Table 6. Hyponyms used only once.

The analysis of building in the CTC is thus consistent with previous studies of the vocabulary of scientific and technical discourse (Cowie 1988; Roldán-Riejos 1999; Coxhead 2000; Coxhead and Nation 2001; Curado-Fuentes 2001; Chung and Nation 2003, 2004; Mudraya 2006). Contrary to what can be expected, the main core of the terminology of a technical corpus is not composed of strictly technical terms. Rather, a high percentage of the vocabulary consists of semi-technical or sub-technical vocabulary, on the one hand, and of polysemous words from general registers, on the other. Although *house* is the most frequent hyponym of *building*, only 33 of its 112 hyponyms are used. Similarly none of the 32 hyponyms of *hospital* (37) is used; only 5 of the 30 hyponyms of *hotel* (126) and 5 of the 33 hyponyms of *restaurant* (10) are found.

Among the 353 WordNet hyponyms (72.28%) which are not used in the CTC corpus are buildings used for keeping animals (e.g. *aviary, cowshed, byre, henhouse, feedlot*) or for working with them (e.g. *abattoir, slaughterhouse, meat house, smokehouse*); also small buildings not meant to be inhabited (e.g. *dog house, doll's house, coal house, toolshed, boat house, woodshed, signal box*). These buildings are not usually designed by architects, mainly because of their auxiliary and secondary purpose. There are clear economic reasons in the design and construction industry and the corpus clearly reflects this particular communicative goal. There also seem to be geographical traits in the corpus and thus certain buildings strongly linked to particular areas of the world are not found in the corpus (e.g. *imarets* from Turkey; British *caffs* and *butteries*; *saltboxes* in New England; Scottish *kirks*; Swiss *chalets*). Cultural reasons might explain the absence of such hyponyms as *whorehouse, brothel, massage parlor, morgue, crematory,* or *shooting gallery.* Slang, colloquial or pejorative terms have also been avoided in the corpus (e.g. *rathskeller, shebeen, gambling den, booby hatch, crazy house, greasy spoon*).

The lexis of construction engineering can be understood as forming a network of relationships "rather than a listing of words as in a published dictionary" (Saeed 1997: 63). The analysis of the hyponyms of the noun *building* in the CTC is followed by its formal representation in a tree-diagram in which semantic relations establish taxonomical hierarchies which classify and categorise disciplinary knowledge. Starting with *entity*, one of the 25 basic semantic primes, and ending with *abbey*, the lowest level of subordination, the resulting word tree would be the following:

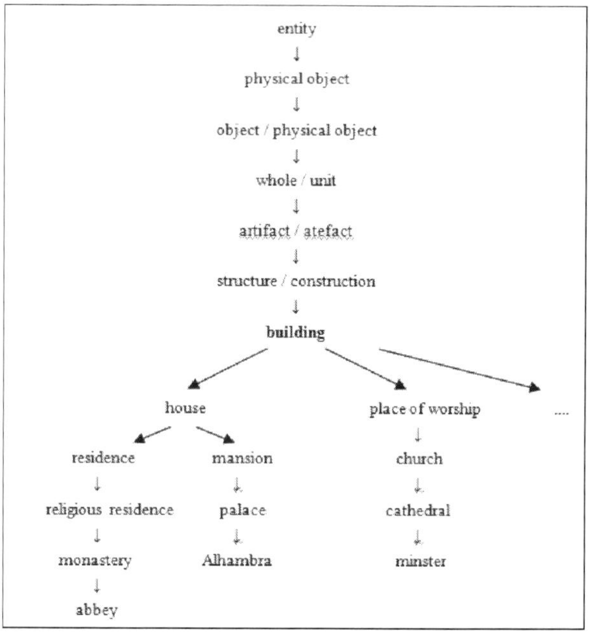

Graph 3. The Word Tree of *building.*

This Word Tree shows the lack of specificity of the CTC: despite the frequency of first-level hyponyms, not many of the hypernym-hyponym chains reach subsequent levels of specificity. Only *house* reaches the fifth level and *place of worship,* the fourth level. The rest of the hypernyms only have hyponyms in two levels of specificity.

The network of hyponyms of the noun *building* is dominated by the predominance of occurrence of the two senses of the noun *house*, followed by the hyponyms of the hypernym *place of worship.* This connects expert knowledge of buildings with the linguistic and discourse realisation of disciplinary knowledge. Construction textbook writers share a specific understanding of what buildings are and what type of buildings are relevant for their profession, which is reflected in their choice of hyponyms and in the clear prevalence of certain types of building: on the one hand, historically and architecturally influential buildings and, on the other hand, houses, the prototypical building.

5. Conclusions

By presenting the analysis of the hyponymy relations in the CTC and the design of the Word Tree of the noun *building*, this paper has attempted to contribute to a better understanding of the taxonomy of buildings which are relevant in the construction industry, as well as to more precisely determine what professionals mean when they say, or write, *building*.

The contrast of the CTC with the lexical database of the WordNet project for the hyponyms of the noun *building* has shown the very low frequency of use of the hyponyms compiled by WordNet. This suggests a sharp distinction between the reality of disciplinary discourse and the encyclopaedic knowledge of the database. Although the database has tried to cover every single sense of every possible building, the frequency analysis of the corpus opposes the reality of the profession, for which a very limited number of buildings, and therefore of the terms which designate them, seem relevant.

Economic, geographical and cultural reasons seem to have influenced the textbook authors' choice of lexis and their preference for commercial and institutional construction. Buildings are thus an economic activity for people or governments linked to a particular geographical and socio-cultural context. The prototypical building in the CTC would be a *house*, a *hotel* or a *centre*. Historical buildings such as *churches*, *castles* and *palaces* are also the topic of research in the professional area of construction engineering.

References

Bloomer, Aileen / Griffiths, Pattrick / Merrison, Andrew J. 2005. *Introducing Language in Use*. London: Routledge.

Carter, Ronald / McCarthy, Michael (eds) 1988. *Vocabulary and Language Teaching*. London: Longman.

Carter, Ronald / Goddard, Angela / Reah, Danuta / Sanger, Keith / Bowring, Maggie 2001. *Working with Texts: A Core Introduction to Language Analysis*. London: Routledge.

Cowie, Anthony P. 1988. Stable and Creative Aspects of Vocabulary Use. In Carter, Ronald / McCarthy, Michael (eds) *Vocabulary and Language Teaching*. London: Longman, 126-137.

Chen, Qi / Ge, Guang-Chun 2007. A Corpus-Based Lexical Study on Frequency and Distribution of Coxhead's AWL Word Families in Medical Research Articles (RAs). *English for Specific Purposes*. 26/4, 502-514.

Chung, Teresa M. / Nation, Paul 2003. Technical Vocabulary in Specialised Texts. *Reading in a Foreign Language*. 15/2, 103-116.

Chung, Teresa M. / Nation, Paul 2004. Identifying Technical Vocabulary. *System*. 32, 251–263.

Coxhead, Averil 2000. A New Academic Word List. *TESOL Quarterly*. 34/2, 213-238.

Coxhead, Averil / Nation, Paul 2001. The Specialised Vocabulary of English for Academic Purposes. In Flowerdew, John (ed.) *Research Perspectives on English for Academic Purposes*. Cambridge: Cambridge University Press, 252–267.

Cruse, Alan D. 1986. *Lexical Semantics*. Cambridge: Cambridge University Press.

Cruse, Alan D. 2000. *Meaning in Language: An Introduction to Semantics and Pragmatics*. Oxford: Oxford University Press.

Cruse, Alan D. 2002. Hyponymy and its Varieties. In Green, Rebecca / Bean, Carol A. / Myaeng, Sung H. (eds) *The Semantics of Relationships: An Interdisciplinary Perspective*. Dordrecht: Kluwer Academic Publishers, 3-22.

Cruse, Alan D. 2003. The Lexicon. In Aronoff, Mark / Rees-Miller, Janie (eds) *The Handbook of Linguistics*. Malden (Mass.): Blackwell Publishers, 238-264.

Cruse, Alan D. 2006. *A Glossary of Semantics and Pragmatics*. Edinburgh: Edinburgh University Press.

Curado-Fuentes, Alejandro 2001. Lexical Behaviour in Academic and Technical Corpora: Implications for ESP Development. *Language Learning & Technology*. 5/3, 106-129.

Darian, Steven 2003. *Understanding the Language of Science*. Austin: University of Texas Press.

Fairclough, Norman 2003. *Analysing Discourse: Textual Analysis for Social Research*. London, New York: Routledge.

Fellbaum, Christiane (ed.) 1998. *WordNet. An Electronic Lexical Database*. Cambridge, MA: MIT Press.

Gregory, Howard 1999. *Semantics*. London: Routledge.

Griffiths, Patrick 2006. *An Introduction to English Semantics and Pragmatics*. Edinburgh: Edinburgh University Press.

Hudson, Richard A. 1995. *Word Meaning*. London: Routledge.

Jackson, Howard / Amvela, Étienne Z. 2000. *Words, Meanings and Vocabulary: An Introduction to Modern English Lexicology*. London: Cassell.

Kearns, Kate. 2006. Lexical Semantics. In Aarts, Bass / McMahon, April (eds) *The Handbook of English Linguistics*. Malden, MA: Blackwell Publishers, 557-580.

Kempson, Ruth M. 1977. *Semantic Theory*. Cambridge: Cambridge University Press.

Kreidler, Charles W. 1998. *Introducing English Semantics*. London: Routledge.

Lyons, John 1968. *Introduction to Theoretical Linguistics*. Cambridge: Cambridge University Press.

Lyons, John 1977. *Semantics*. Cambridge: Cambridge University Press.

Lyons, John 1995. *Linguistic Semantics: An Introduction*. Cambridge: Cambridge University Press.

Miller, George 1998. Nouns in WordNet. In Fellbaum, Christiane (ed.) *WordNet. An Electronic Lexical Database*. Cambridge, MA: MIT Press, 235-244.

Mudraya, Olga 2006. Engineering English: A Lexical Frequency Instructional Model. *English for Specific Purposes*. 25/2, 235-256.

Nation, Paul / Kyongho, Hwang 1995. Where Would General Service Vocabulary Stop and Special Purposes Vocabulary Begin? *System*. 23/1, 35-41.

Orna-Montesinos, Concepción 2008. A Contribution to the Lexis of Construction Engineering Textbooks: The Case of *Building* and *Construction*. *Ibérica*. 16, 59-79.

Roldán-Riejos, Ana M. 1999. Applications of Cognitive Theory to Interdisciplinary Work in Languages for Specific Purposes. *Ibérica*. 1, 29-37

Saeed, John I. 1997. *Semantics*. Oxford: Blackwell.

Salkie, Raphael 1995. *Text and Discourse Analysis*. London: Routledge.

Thurstun, Jennifer / Candlin, Christopher 1998. Concordancing and the Teaching of the Vocabulary of Academic English. *English for Specific Purposes*. 17/3, 267-280.

Trimble, Louis 1985. *English for Science and Technology. A Discourse Approach*. Cambridge: Cambridge University Press.

Widdowson, Henry G. 1998. *Linguistics*. Oxford: Oxford University Press.

Yule, George 2001. *The Study of Language*. Cambridge: Cambridge University Press.

Appendix

- apartment building (13)
  - tenement (1)
  - walk-up (1)
- architecture (1)
- bathhouse (3)
- bowling alley (1)
- center/centre (124)
  - call centre (1)
  - community center (3)
  - conference center / centre (14)
  - research center (1)
- clubhouse (1) , club (4)
- dormitory (2) , hall (2)
- hall of residence (2)
- farm building (2)
  - barn (3)
  - stable (1), stalls (1)
- gambling house*
  - casino (1)
- summerhouse (1)
- government building (2)
  - capitol (2)
  - courthouse (5)
  - embassy (4)
- greenhouse (35)
  - conservatory (4)
- hall (68)
  - city hall (5)
- hall (5)
- hotel (126)
  - hostel (2), inn (2), lodge (2)
  - spa (1)
  - resort (6)

- casino-hotel (2)
- house (419)
    - bungalow (2), cottage (7)
    - cabin *
        - log cabin (1)
    - country house (2)
        - summer house (1)
        - villa (a) (8)
        - villa (b) (1)
    - detached house (7)
    - farmhouse (2)
    - guesthouse (4)
    - lodge (1)
    - maisonette (4)
    - mansion (4)
        - palace (17), castle (13)
        - stately home (3)
    - residence (3)
        - palace (1)
            - Alhambra (2)
            - Versailles (4)
        - religious residence*
            - convent (2)
            - monastery (3)
                - abbey (5)
        - Vatican (2)
    - town house (2)
        - terraced house (11)
    - solar house (23)
- house (12)
    - bathhouse (3)
    - courthouse (6)
    - sports arena (10)
    - icehouse (5)
    - pumping station (12)
- library (14)

- medical building (1), health facility (2)
  - clinic (4)
  - hospital (37), infirmary (1)
- funeral parlor (1)
- observatory (2)
- office building (20), office block (4)
- outbuilding (1)
  - garage (21)
  - shed (10)
- place of worship (3), house of worship (1)
  - chapel (11)
    - Sistine Chapel (2)
  - church, (133), church building (5)
    - abbey (6)
    - basilica (9)
    - cathedral (26)
      - minster (2)
      - Chartres Cathedral (4)
    - cathedral (15)
  - mosque (7)
  - shrine (2)
  - temple (35), tabernacle (1)
  - temple (62)
    - pantheon (5)
    - Parthenon (21)
- planetarium (1)
- restaurant (10)
  - brasserie (1)
  - cafe (16), coffeehouse (1)
  - cafeteria (1)
  - lunchroom (2)
- Roman building (3)
  - basilica (3), Roman basilica (1)

- ruin (19)
- school (52)
  - conservatory (1)
- skyscraper (18)
- student union (1)
- tavern*
  - public house (1), pub (6)
- temple (3)
  - ziggurat (3)
- theatre/theatre (66)
  - theater in the round (1)
  - cinema (6)
  - music hall (1)
  - opera (7), opera house (23)
- Houses of Parliament (1)

Table 7. The word tree of building.

Nuria Edo

The Creation of an Active, Corpus-Based Dictionary on Industrial Ceramics from Specialised Lexicography

1. Introduction: theoretical grounds for entry generation

The main goal of this paper is to offer a significant and illustrative part of the results obtained in an ongoing project aimed at the creation of an active, bilingual (English-Spanish, Spanish-English) dictionary of the ceramics industry. As the entries selected will show, the basic objective of this specialised dictionary is to systematise, within the framework of contrastive lexicography, the terminological units of the ceramics industry field in a useful and realistic manner, emphasising the communicative and associative nature of such units.

It is necessary, however, to first contextualise the research presented here, which is based on three main theoretical grounds:

- the communicative nature of terminology, drawing on Cabré's Communicative Theory of Terminology or CTT (1999);
- the convenience of creating active dictionaries capable of surpassing the mere terminological approach to become works of specialised lexicography;
- the need of using corpora that allow the prospective dictionary users to have direct access to real, accurate and high-quality examples of use.

Regarding the first issue, the terminological perspective defined by Cabré (1999) plays a leading role due to the close and realistic view it provides of the terminological reality. This author emphasises the need for materials which satisfy the communicative demands in speciality areas and which ultimately respond to the expectations of a constantly developing and changing society that opts, often out of pure necessity, for more pragmatic and applied approaches. Therefore, as its own name indicates, the CTT accounts for the communicative nature of terminology and its interdisciplinarity, and integrates the most theoretical aspects of terminology with a coherent practice of it; it also advocates the treatment of terminology and specialised languages from an integrating perspective, based on the idea of *continuum* rather than on abrupt and artificial divisions that do not reflect the linguistic reality nor the needs posed. In short, the CTT approaches terminology from a natural perspective, enhancing its active, communicative dimension.

With regard to the second theoretical pillar, that is, the convenience of making dictionaries from an active approach[1], a special emphasis has been placed over the last years on the creation of dictionaries that satisfy the specific needs of the users for whom they are intended (Kromann/Riiber/Rosbach 1991). In this sense, Cowie (1979, in Kromann/Riiber/Rosbach 1991) pointed out that the reach, organisation and metalanguage of dictionaries should analyse and reflect the referential needs of prospective users in the most reliable way possible. Therefore, it seems clear that a key issue when developing a project with the features and approaches above-mentioned is to consider how these needs are going to be met in the specialised lexicographical work, so that decisions regarding monolingualism vs. bilingualism, description vs. prescription, etc. are dealt with in an adequate way. This is precisely what my research intends to do: to fill a gap in specialised communication within the ceramics industrial field in an attempt to contribute to the improvement of expert communication from a terminological point of view.

As the discipline which deals with the compilation, description, management and presentation of terms belonging to speciality domains in one or more languages, terminology – and its more applied branch, terminography – is not a practical activity that can be justified by itself but within a wider context, in this case the ceramics industrial sector. For the development of the prospective dictionary here presented, terminology has been conceived as a field of knowledge aimed at the resolution of real social needs and linked to the optimisation of communication

1 Section 2 explains in detail the kind of data that active dictionaries normally contain in order to be considered so.

among professionals or experts. Thus, the terminological approach adopted here transcends the 'mere' terminographical task and fits better into the label 'specialised lexicography', mainly because of its intended user-friendly, active and communicative features. Moreover, this positioning offers a much more integrating view of the objects of study, according to which every linguistic unit, independently from their degree of specialisation, forms part of the lexicon of the language; thus, lexicology would include any terminological study. As Adelstein (2004) concedes, this is because, from a pragmatic point of view, terms and words are only distinguished by the activation of a series of traits that bestow a specialised character to them.

This view is the basis for the *Principio del valor terminológico*[2] (Cabré 2008), deeply grounded in the need of studying terms in vivo and, according to which, lexical units (LUs) are not words or terms *per se* but units that activate (or not) their terminological value depending on their use in a particular communicative context. Accordingly, the LU is the base unit that may activate a series of speciality traits, depending on the context. In case these traits are not activated, units will be plain LUs (commonly known as words) whereas, in case they do, units will become TUs (commonly known as terms) and both of them can be approached from linguistics. Hence, it may be claimed that terminology is specialised lexicology or a lexicology of the specialised lexical units, which points at the convenience of using the label 'specialised lexicography' to refer to the linguistic practice undertaken here. Consequently, terminology has been conceived in this research as the 'mere' study of terms possible thanks to the activity of compiling and describing them, called terminography. Terminography, at the same time, may be complemented by the wider projection of specialised lexicography for the elaboration of specialised dictionaries oriented to the user and user-friendly, as this study intends.

Finally, a corpus-based approach seems to be the best (if not the only) approach to achieve these goals. By nature, a corpus must be representative and contain real samples produced by competent speakers with a real, proved knowledge of the domain under study, apart from being balanced, equilibrated and with an adequate size.

2 'Principle of terminological value'.

Accordingly, Ahmad/Davies (1992: 22) summarise the potential of this kind of approach as follows:

> A corpus-based approach enables the terminologist to identify, study and record terms in their natural habitat, and translators, as users of the term bank, to have customised access to a full set of information about all the terms and related terminological data of a domain.

Likewise, a corpus-based study provides evidence of the different meanings a lexical unit may have depending on the context in which it is given. Moreover, the numerous computer tools developed in the field of terminotics allow the terminographer to undertake a complete and detailed analysis of different aspects of corpora ranging from quantitative to purely linguistic aspects, this being an advantage to take into much consideration.

Following Cabré's (2008) *Principio del valor terminológico,* graph 1 offers a very simple example from the domain analysed. We can observe the difference between the first use of the lexical unit *orange peel*, a general one, and the second use, in which context activates specialised traits. In its second instance, *orange peel* is a TU forming part of the prospective dictionary.

SPECIALISED USE: *A vibrant* ***orange peel*** *ceramic glaze graces this distinctive table lamp design* *Salt glaze has the texture of* ***orange peel****, formed on stoneware by throwing common salt into the kiln at the peak temperature.* NON-SPECIALISED USE: *Add* ***orange peel****, orange juice, brown sugar, butter, salt and cinnamon.* *Top each serving with whipped cream; sprinkle with* ***orange peel****, if desired*

Graph 1. Specialised and non-specialised use of the LU *orange peel* grounded on a corpus-based analysis.

Apart from real contexts of use and data of collocational nature, the analysis of corpora may provide statistical and frequency data that complement purely linguistic ones. Hence, the consideration of all these issues, together with the in-depth analysis of the context of use of the specialised bilingual tool to develop (needs analysis), allow the terminographer to create a dictionary aimed at satisfying the specific demands of the ceramics industrial domain and providing responses to 'what is needed' in the sector from the terminological-communicative point of view. Therefore, the theoretical positioning adopted and the way of elaborating the entries presented in the next section is giving way to an active, user-friendly dictionary that faithfully tries to reflect the terminological reality under analysis. At the same time, the design of the entries intends to meet the expectations and needs of prospective users as the label 'specialised lexicography' implies.

2. Presentation of selected final entries from the dictionary

The elaboration of active entries is not a simple task and, predictably, will not always satisfy the expectations of all prospective users, at least in the same level or way. Nonetheless, in this work of specialised lexicography all factors involved in dictionary-making have been taken into account with the ultimate aim of generating a coherent and useful dictionary.

In general, graphs 2 and 3 show what could be considered two model entries from the specialised dictionary on industrial ceramics: these entries have been designed and generated with the final objective of gathering all the information necessary for the correct understanding of the concept on the part of the user, and for the correct production on the part of the speaker. Active dictionaries, as opposed to passive[3] ones, incorporate more information aimed at enhancing correct production.

Therefore, as an active approach requires, these bilingual entries contain up to ten pieces of information based on Posteguillo/Piqué-Angordans' (2005) research:

---

3 Normally, what is called here a bilingual passive specialised dictionary (in opposition to active ones) contains fewer data, namely the translation into the target language –accompanied at times by some technical information– and vice versa.

- The equivalents in L2.
- The lexical and grammatical category.
- The semantic field to which the specialised lexical unit belongs.
- Cross references.
- Abbreviations.
- Real contexts of use.
- Definitions and notes on use (including in many cases the translation between hyphens of many terms that may turn out to be problematic or opaque in meaning).
- Synonyms.
- Collocations, both within the contexts or definitions as such or in the form of polylexematic/polylexical terms acting as subentries with the lemma acting as the base term.
- The different meanings that homonymic or polysemic TUs may have, differentiating them by means of superindexes.
- The spelling or ortographic variants of a term.

The way of including these pieces of information in specific active entries from our prospective dictionary is illustrated in graphs 2 and 3 below:

arc *n*: GRAL arco de circunferencia; línea perteneciente a la circunferencia que une dos puntos de ella. [Exp: **arc (electric)** (QUIM arco (eléctrico); descarga eléctrica luminosa entre dos electrodos en el seno de un gas que se ioniza), **arc furnace** (ELABPROC horno (eléctrico) de arco; horno en el que el calor es generado por un arco eléctrico que salta entre dos electrodos o entre un electrodo y la propia materia ◊ *Fused alumina made in the electric arc furnace is usually crushed, and sieved into a range of grit sizes for use as an abrasive; V. electric furnace; Heroult electric arc furnace; direct-arc air furnace),* **arc ion plating, AIP** (ELABPROC recubrimiento iónico por arco ◊ *Some characteristics of the arc ion plating technology are high ionization and high deposit rate since the coatings deposited by this method are characterizedby high combination, adhesion, density and coherency etc.),* **arcade** (CONST arcada, soportal/soportales; pórtico, a veces con arcos, que tienen algunos edificios/casas en sus fachadas y delante de las puertas y tiendas que hay en ellas), **arcaded** (CONST arcos, formado por, arcos, decorado con), **arc-furnace roof brick** (INDAFIN ladrillo de bóveda de horno eléctrico), **arc-image furnace** (MAQELAB horno de arco de imagen ◊ *An Arc-Image furnace can heat any samples rapidly to a very high temperature in a clean condition, without containers or crucibles, in any oxidizing, inert or reducing atmospheres; V. image furnace),* **arc-spraying** (ELABPROC proyección arco ◊ *Arc spraying is the highest productivity thermal spraying process; V. plasma spraying)].*

Graph 2. Model entry *arc* including subentries and information for the active use of the term.

Ash1 *n/v:* SUBPROD/ESMREL ceniza, resíduos de combustión; resíduos inquemados; residuo incombustible resultante tras la combustión/calcinación de una sustancia ◊ *Ash may be used in combination with coloring oxides and other glaze materials to produce a wide range of surface effects and colors in glaze*; V. *volcanic ash.* [Exp: **ash2** (ELABPROC calcinar, incinerar; reducir a cenizas), **ash cump** (MAQELAB depósito de cenizas), **ash glaze** (ESMREL esmalte de ceniza; esmalte que utiliza ceniza orgánica (papel, madera) como fuente de óxidos (sílice, aluminio, oda, etc.) y que a menudo presenta una textura moteada -*mottled*- o entreverada -*streaky*- dependiendo de la cantidad de ceniza empleada ◊ *A normal glaze application fired at cone 10 results in a glaze similar to an ash glaze, with streaks)*, **ash-glazed** (ESMREL esmalte/esmaltado de ceniza ◊ *A typical ash-glazed surface is characterized by rich, fluid streaks of texture on the surface that are left as the ash melts*; V. *ash glaze*), **ashing** (ELABPROC calcinación, incineración; combustión ◊ *Minerals lose water upon ashing*), **ash-pan** (MAQELAB V. *ash-pit*), **ash-pit** (MAQELAB cenicero, cenizal (caldera); parte del horno en la que cae la ceniza ◊ *As the firing progressed the ash pit began to fill and eventually filled almost to the point of obstructing the arch into the ware area of the kiln*), **ashtray** (GRAL cenicero; sitio donde se recoge o echa la ceniza)].

Graph 3. Model entry *ash* including subentries and information for the active use of the term.

The inclusion of this kind of data in a specialised dictionary entry, which at first sight might seem rather obvious, is not so, if you compare the specialised entry with entries from previous lexicographical works (see graph 4). It must be said, however, that although these works do not constitute samples of a communicative, active and specialised lexicographical approach to terminology, they prove very useful to become familiar with terms and to find equivalents in other languages. By comparing graphs 2, 3 and 4, it may be easily noticed what an active approach can provide.

Oven. Obsolete term for a kiln, more particularly the old type of BOTTLE-OVEN (q.v.).

From: Murfin, D. (ed.) (1994): **Dictionary of Cera mics Arthur Dodd**, *London. The Institute of Materials.*

FRITAR: Someter al fritado a un esmalte, es decir, hacer una "frita" con él *(Ver)*.

EXFOLIACIÓN: Carácter diagnóstico que permite identificar minerales, según el modo de romperse al golpe, ya sea en láminas (exfoliación basal), en cubos o romboedros, etc. Cuando la rotura es irregular (sin formas fijas), se la denomina "fractura" *(Ver)*.

From: Fernández Chiti, J. (ed.) (1985): **Diccionario de cerámica. Tomo 2,** ***Condorhuasi.***

English	Deutsch	Español
Tunnel dryer	Tunneltrockner, Kanaltrockner	Secadero túnel
Tunnel kiln	Tunnelofen	Horno túnel

From: **Diccionario de azulejos cerámicos CEC (Deutsch-English-Español-Français-Italiano-Svenska).**

Graph 4. Entries from previous lexicographical works showing a passive approach.

These entries intend to be an example of the evolution undergone in the terminographical field, and how it has acquired the dimension of specialised lexicography. More specifically, the information contained in the entries presented in this paper offers important data from the communicative point of view, both for decoding and encoding, with a marked emphasis on production (active use of language) generally disregarded in previous terminographical works. The fact of not including this kind of information could lead to mistakes or at least to a lack of accuracy, so inadvisable when dealing with terminological questions.

For instance, in the polylexematic term *alligator hide*, the base *alligator* has, as its Spanish equivalent, a literal translation: the term *caiman* (see graph 5). However, thanks to corpora and collocations analyses, it may be observed how the subentry or collocation *alligator crusher* does not have as its equivalent in Spanish the term *machacadora de caiman*, as could be expected if merely literal translations were to be considered, but *machacadora de mandíbulas*[4]. Likewise, *alligator's teeth* is not translated as *dientes de caimán* but as *dientes de cocodrilo*[5]. This kind of knowledge – that is, the fact that one says *dientes de cocodrilo* instead of *dientes de caiman* in the target language – is what differentiates the expert from the layman, and a person who accurately transmits knowledge in its own field of work (even in a language different from its mother tongue) from another who does not. This knowledge is what distinguishes an active, communicative approach from a passive one.

alligator n: GRAL caimán. [Exp: **alligator crusher** (MAQTRANS machacadora de mandíbulas; machacadora en la cual el material se introduce por la parte superior y entra en una cámara que contiene la mandíbula encargada de presionarlo contra la pared de dicha cámara con una fuerza extrema, triturándolo en trozos más pequeños), **alligator hide** (ESMREL piel de caimán, piel de naranja; defecto en la superficie del esmalte que constituye un caso severo de la piel de naranja -orange peel- y que puede ser debido a un secado demasiado rápido o a una aplicación demasiado gruesa de esmalte; puede tener también fines decorativos ◊ *Alligator hide appears when the glaze does not flow uniformly over the surface of the ware*; V. *orange peel*), **alligators' teeth** (CALIDAD dientes de cocodrilo (en la pieza extruída))].

Graph 5. Entry term *alligator* with its subentries.

Lexicalised plurals, that is, entries of the *bacteria* type (see graph 6), are indicated in their corresponding entry by the grammatical category *n. pl.* and the note of use 'forma plural del sustantivo'.

4 Mandíbulas (ES); Jaws (EN)
5 Cocodrilo (ES); Crocodile (EN)

bacteria n pl: GRAL bacterias; (forma plural del sustantivo) microorganismo unicelular procariota causante de las fermentaciones, enfermedades o putrefacción en los seres vivos o en las materias orgánicas ◊ *Although they are lead-free and non-toxic, they are not dinnerware safe because the rough surface may tend to trap bacteria.* [Exp: **bacterial** (GRAL bacteriano, bacterial, bacteriológico ◊ *For the tile, washing and rinsing reduced the bacterial counts by about 900 to 1*; V. *anti-bacterial/antibacterial*), **bacterial corrosion** (MANTNMTO corrosión bacteriana; en un material, destrucción producida por los procesos químicos originados por la actividad de ciertas bacterias que generan sustancias como el sulfuro de hidrógeno, amoníaco y ácido sulfúrico), **bacterial leaching** (GEO-MIN-CRIST lixiviación bacteriana, biolixiviación; proceso consistente en la acción de bacterias sobre los minerales para acelerar la reacción de los minerales sulfurados)].

Graph 6. Entry term *bacteria* illustrating the treatment of lexicalised plurals.

Likewise, the entry term *average* in graph 7 illustrates how a specialised lexicographical approach also highlights the way to construct useful and common expressions. For instance, in the entry *average* we find the expressions 'superior/inferior a la media' and 'por término medio' – both of which may be useful, given the domain under study, for encoding purposes. In the same way, graph 8 illustrates the use of the verb *agree* – so necessary in commercial activities – both in its transitive and prepositional variants.

average[1] *a/n/v*: GRAL promedio, media aritmética; referido a números, la suma de todos ellos dividida entre el número de sumandos; superior/inferior a la media: *above/below average*; por término medio: *on (the) average*. [Exp: **average**[2] (GRAL medio; que presenta las características o condiciones generales de un grupo ◊ *Studies of such systems have shown that where crystallite growth is the predominant process, the average pore size increases*), **average**[3] (GRAL promediar; calcular la media de algo), **average consumer** (GRAL/COM consumidor medio ◊ *The average consumer has an idea that they want to use a tile or stone product.*), **average dip** (GEO-MIN-CRIST/EXPL-CANT buzamiento medio (geología); inclinación media de un filón o capa del terreno ◊ *The quarrying operations are carried out taking advantage of the layering of the stone, almost flat and horizontal (average dip of 2° toward south)*), **average hade** (GEO-MIN-CRIST/EXPL-CANT pendiente media (geología); inclinación media ◊ *The fault has an average hade to the west of about 30 degrees*; V. *average dip*), **average sample** (CALIDAD muestra media, muestra promedio ◊ *An "average" sample is taken, rather than looking at still samples, which may vary from one sample to the other*), **average value** (GRAL valor medio/promedio V. average), **awash** (GRAL inundado por; cubierto de agua o, en sentido figurado, de algo que recuerda al agua ◊ *The bath is awash in aquamarine tile on both the floor and the walls*)].

Graph 7. Entry term *average*.

agree v: GRAL acordar, estar/ponerse de acuerdo; convenir, acceder, aceptar, aprobar, concertar, concordar, consentir, pactar; el verbo agree puede ser transitivo *-the two companies have agreed terms-*, preposicional con on *-they agreed on the price and a bargain was struck-*, preposicional con to *-the company agreed to pay Paul 500-*, preposicional con with *-he agrees with the sales manager*. [Exp: **agreed** (COM/GESTIÓN acordado, concertado, conforme, convenido ◊ *When an order is placed, items such as size, thickness, type of surface, colour relief, abrasion class for glazed tiles and any special properties such as chemical resistance shall be agreed by the parties concerned*), **agreed price** (COM precio acordado/concertado ◊ *Unless otherwise agreed in writing, the agreed price shall be fixed and not index-adjusted and shall also include packaging of goods*), **agreed upon, as may be** (COM según se convenga ◊ *After the expiry of twelve months, any balances outstanding will be settled in such a way as may be agreed upon between the two parties*), **agreement** (GRAL/GESTIÓN acuerdo; contrato), **agreement with, in** (GRAL/GESTIÓN conforme a; de acuerdo con ◊ *Each arrangement shall be made in agreement with the Agency*), **agreement, by** (COM a convenir ◊ *The setting of prices by agreement among producers and distributors; V. negotiable*)].

Graph 8. Entry term *agree*.

The entry *abroad* (graph 9) also illustrates the commonest usages of the term in the field, and offers a frequent construction: 'irse al extranjero'.

abroad adv: GRAL extranjero (en el); to go abroad: ir-se al extranjero ◊ *We specialize in manufacturing, processing and trading abroad ceramic tiles.*

Graph 9. Entry term *abroad*.

In addition to this, the kind of entries generated also show the natural associations of the terms, that is, their collocational nature. For example, graphs 10 and 11 show how *ban* is frequently accompanied by 'use' in the domain of industrial ceramics, giving way to expressions such as 'banned the use' ('prohibió el uso'). These data (collocations), contained both in the subentries as such and/or in the contexts of use, show the associative nature of terms, that is, their tendency to appear accompanied by some lexical units more frequently than with others.

ban *v/n:* GRAL prohibir; especialmente empleado cuando se prohíbe algo por vía legal ◊ *Arizona banned the use of lead-based glaze in nursing homes*. [Exp: **ban** (GRAL prohibición ◊ *The Federal Asbestos Ban did not go into effect until 1989*)].

Graph 10. Entry term *ban*.

Likewise, the entry in graph 11 – and more specifically its context – highlights the tendency of the adjective *banded* (a subentry of the entry term *band*) to appear with 'clay layers'. They naturally collocate because they tend to appear together or in a close position.

banded (PROPMATPRI/GEO-MIN-CRIST bandeado; dícese de algo con capas alternas casi paralelas que se diferencian en el color o la composición mineral *-mineral composition-*, por lo que muestran una alternancia de bandas *-alternate bands-* en la sección transversal *-cross section-* ◊ *Banded clay layers are suggestive of distinct pulses of rapid mud deposition*)

Graph 11. Entry term *banded.*

Other entries, such as the one in graph 12, show notes on the use or provide implicit shades of meaning in the very same concept. In this case, the polylexical term *acceptance of goods* implies ('implica') 'disposición a abonar su importe'.

acceptance of goods (DISTRIB aceptación de mercancías; implica "disposición a abonar su importe" ◊ *Acceptance of goods occurs when the buyer, after a reasonable opportunity to inspect the goods, signifies to the seller that the goods are conforming or that he will take or retain them*; V. *brand acceptance*)

Graph 12. Entry term *acceptance of goods.*

Many notes on the use make reference as well to variants depending on the geographic area in which they are given (see graph 13).

air-setting refractory cement (INDAFIN cemento refractario de fraguado al aire; en EEUU, el cemento refractario de fraguado al aire se conoce como "*air-setting refractory mortar*" ◊ *An air setting refractory cement was mixed with water to yield a paste which was formed into a 1-inch cube*; V. *chemically-bonded refractory cement*)

Graph 13. Entry term *air-setting refractory cement.*

Moreover, considering the economy principle, many entries such as the one in graph 14 do not include definitions because they already incorporate explanatory contexts in which terms are explained.

actual market value (COM valor actual de mercado ◊ *The closer your price is to actual market value -what a knowledgeable seller will accept and a knowledgeable buyer will pay- the sooner it will sell*)

Graph 14. Entry term *actual market value*.

Finally, the inclusion of the semantic field in the entries is illustrated in graph 15 by the word PROPMATPRI. This kind of words in capital letters constitute abbreviations for the different fields and subfields in which the domain has been previously structured by the terminographer, giving way to a *field diagram*. A *field diagram* is an organising and organised structure showing the branches that integrate a speciality domain and the ones that are directly related to it. The idea is that every TU belonging to a speciality domain should be able to be meaningfully placed in one of these branches or semantic fields. Accordingly, a semantic field is a group of notions that can be grouped under the heading of a key notion (Auger/Rousseau 1987).

This is what PROPMATPRI in graph 15 tries to convey: that the term *vesicular* belongs to the semantic field 'propiedades de las materias primas' (what in English has been called RAWMATPROP 'raw materials properties'). In industrial ceramics, *vesicular* presents as its key notion the fact of being a property, a quality of raw materials, and the inclusion of this piece of information in the entry might be very beneficial for its correct understanding on the part of the user.

vesicular *a:* PROPMATPRI vesicular; sólido que contiene gran cantidad de pequeñas burbujas ◊ *A bloated clay may be vesicular.*

Graph 15. Entry term *vesicular*.

Dictionaries with entries such as the one presented in this study basically aim at satisfying the needs of prospective users. The needs analysis conducted for this research supported the decision to take both a specialised lexicographical approach and an active perspective, since they both foster the success of the communicative exchange by enhancing a satisfactory understanding and production on the part of the user and by analysing terminology *in vivo*, as it is used in real life situations.

3. Conclusion

The basic objective of the prospective specialised dictionary of the ceramics industry is, first of all, to systematise the terminological units of this domain within contrastive specialised lexicography (English-Spanish, Spanish-English). Thus, this paper has been mainly aimed at showing a series of results in the form of final dictionary entries that represent or illustrate in a smaller but representative scale what the dictionary will be like. Thanks to the active, communicative and corpus-based approach adopted, this prospective dictionary will predictably allow the user to find, not only the equivalents of the terms in the other languages of work, but also illustrative examples of the term in use, explanations and definitions of the term, notes on the use, collocations, cross references, the semantic field of the terms, synonyms, acronyms, abbreviations, and other relevant information helping to understand the concept and showing how to use the term. Each entry will include up to 10 different types of pieces of information – technical, grammatical and pragmatic –, considering always the prospective users of the dictionary and their needs. Accordingly, what tries to differentiate this dictionary from other previous undertakings of the sort or simply from other specialised dictionaries, is the pragmatic, communicative perspective from which it is being generated and which is shaped in the elaboration of an active dictionary very much concerned with enhancing correct understanding and (specially) production on the part of the speaker.

A crucial issue in this paper has been the idea of creating a specialised dictionary within specialised lexicography, that is, basically grounded on communication and which departs from the intrinsic value of vocabulary in the communicative process, from the study of its active and adequate usage and from the fundamental role played by situations of use within a discourse community.

Likewise, the prospective dictionary is intended to fit the pertinence, clarity and economy criteria, so necessary for the lexicographical task in general: the pertinence criteria will determine whether the lexical units can be inserted within the field diagram and thus belong to the domain under study, i.e. the ceramics industry; the clarity criterion is basic for the functionality of the final product and is reflected, for instance, in the enumeration by means of superindexes of the different meanings a terminological unit may have; finally, the economy criterion will determine, for instance, the exclusion from the dictionary of both sporadic coinings and those consolidated ones but so transparent from a semantic point of view that their meaning is clearly deductible from the meaning of the words that are part of it.

In order to achieve all this, this paper has used as its mainstay the terminological perspective defined by Cabré (1999) in her CTT, which emphasises the necessity to contribute to the field of terminology with materials that satisfy the communicative demands and expectations of the discourse community. Therefore, a work like this tries to lighten the socio-communicative demands of the ceramics industrial sector by fostering competitiveness in terms of linguistic and communicative competence both in English and Spanish.

References

Adelstein, Andreína 2004. *Unidad léxica y valor especializado. Estado de la cuestión y observaciones sobre su representación*. Barcelona: Institut Universitari de Lingüística Aplicada, Universitat Pompeu Fabra. Serie Tesis 5. <http://www.iula.upf.edu/publi036.htm>

Ahmad, Khurshid / Davies, Andrea 1992. *Knowledge Processing 7: Methods of Corpus Linguistics and Terminology Extraction*. Department of Mathematical and Computing Sciences, University of Surrey, CS-92-10.

Auger, Pierre / Rousseau, Louis 1987. *Metodologia de la recerca terminològica*. Sant Adrià de Besòs: Departament de Cultura de la Generalitat de Catalunya.

Cabré Castellví, María Teresa 1999. *La terminología, representación y comunicación: elementos para una teoría de base comunicativa y otros artículos*. Barcelona: Institut Universitari de Lingüística Aplicada (IULA), Universitat Pompeu Fabra.

Cabré Castellví, María Teresa 2008. El principio de poliedricidad: la articulación de lo discursivo, lo cognitivo y lo lingüístico en Terminología (I). *Ibérica*. 16, 9-36.

Diccionario de azulejos cerámicos CEC (Deutsch-English-Español-Français-Italiano-Svenska). Faenza Editrice.

Fernández Chiti, Jorge (ed.) 1985. *Diccionario de cerámica*. Tomo 2, Condorhuasi.

Kromann, Hans-Peder / Riiber, Theis / Rosbach, Paul 1991. Principles of Bilingual Lexicography. In Hausmann, F. J. / Reichmann, O. / Wiegand, H. E. / Zgusta, L.E. (eds) *Wörterbücher/Dictionaries/ Dictionnaires. Ein internationales Handbuch zur Lexikographie / An International Encyclopedia of Lexicography / Enciclopédie internationale de lexicographie 3*. Berlin-New York: De Gruyter, 2711-2728.

Murfin, David (ed.) 1994. *Dictionary of Ceramics Arthur Dodd*. London: The Institute of Materials.

Wüster, Eugen 1979. *Einführung in die allgemanie Terminologielehre und terminologische Lexikographie*. Wien: Springer (Spanish version directed by Cabré Castellví, María Teresa 1998. *Introducción a la teoría de la terminología y a la lexicografía terminológica*. Barcelona: IULA, Universitat Pompeu Fabra).

Inés Lareo

New Trends Exploring the Language of Science: The *Corpus of English Texts on Astronomy* (CETA) and Its Tool (CCT) in the Context of the *Coruña Corpus*[1]

1. Introduction

The group of *Multidimensional Studies in English* (MuStE) of the University of A Coruña (www.udc.es/grupos/muste) is carrying out the project *Coruña Corpus: a collection of samples for the historical study of English scientific writing.* Our aim is to select at least six scientific disciplines, such as Astronomy, Philosophy, Life Sciences, Physics, History and Linguistics, containing around three million words written between 1700-1900. The idea is to include an equivalent number of disciplines belonging to Natural Sciences and Humanities. The corpus will allow the diachronic and synchronic study of Modern English scientific English as well as offering a view of the evolution of science. We hope, thus, to offer the scientific and linguistic community a useful tool to continue the study of the evolution of scientific English.

The starting point of the selection of disciplines to compile each sub-corpus has been the UNESCO classification for Science and Technology published in 1978 (table 1). However, since the limits of the different fields have changed across the centuries, the different sub-corpora contain some samples dealing with topics that have been considered "scientific" at that time. For instance, in the case of Astronomy, we have included some works dealing with Astrology,

1 The research here reported has been funded by Consellería de Educación e Coordenación Universitaria (2007/000145-0), Xunta de Galicia (PGIDIT07PXIB104160PR), Ministerio de Ciencia e Innovación (FFI2008-01649/FILO) and Deputación da Coruña. These grants are hereby gratefully acknowledged.

and in the case of Physics, some texts dealing with Alchemy, whose "scientific methods" can be questionable nowadays. The Life Sciences sub-corpus will comprise the fields Natural History and the disciplines of Natural Sciences italicised in Table 1. Our aim is that the *Coruña Corpus* (CC) encompasses a counterbalanced selection of disciplines dealing with topics included in the first field, Natural Sciences, and in the last one, Humanities.

I. Natural Sciences. Astronomy, bacteriology, biochemistry, *biology*, *botanics*, chemistry, *entomology*, geology, geophysics, mathematics, meteorology, mineralogy, computing, physical geography, physics, *zoology* and other allied subjects. II. Engineering and Technology. Engineering sciences such as: chemical, civil, electrical and mechanical engineering and their specialised subdivisions; forest products; applied sciences such as geodesy, industrial chemistry, etc.; architecture; the science and technology of food production; specialised technologies of interdisciplinary fields, e.g. systems analysis, metallurgy, mining, textile technology and other allied subjects. III. Medical Sciences. Anatomy, stomatology, basic medicine, paedriatics, obstetrics, optometry, osteopathy, pharmacy, physiotherapy, public health services, technical health assistance and other allied subjects. IV. Agricultural Sciences. Agronomy, zootechnics, fisheries, forestry, horticulture, veterinary medicine and other allied subjects. V. Social Sciences. Anthropology (social and cultural) and ethnology, demography, geography (human, economic and social), law, linguistics, management, political sciences, psychology, sociology, organisation and methods, miscellaneous social sciences and interdisciplinary, methodological and historical S&T activities relating to subjects in this group. Physical anthropology, physical geography and psychophysiology are normally classified with the natural sciences.

VI. Humanities. Arts (history of art and art criticism, excluding artistic 'research'), ancient and modern languages and literatures, philosophy (including the history of science and technology), prehistory and history, together with auxiliary historical disciplines such as archaeology, numismatics, paleography, genealogy, etc.), religion, other subjects and humanistic branches as well as other methodological and historical S&T activities relating to the subjects in this group.

Table 1. Fields of Science and Technology (*International Standardisation of Statistics on Science and Technology*).

The first sub-corpus compiled is the *Corpus of English Texts on Astronomy* (CETA) published in 2009 to commemorate the International Year of Astronomy coinciding with the 400th anniversary of Galileo's use of a telescope to study the night sky and the anniversary of many other important dates in the history of science, such as the publishing of Kepler's *Astronomia Nova* or Huygens' *Systema Saturnium.*

2. General selection criteria

All the disciplines included in the CC will follow the same selection criteria concerning time-span, authors, texts, samples, number of words, etc. (Moskowich/Crespo 2007; Lareo 2008, 2009a; Lareo/Esteve 2008). The explanations that follow, however, will be focussed on CETA, the sub-corpus presented here.

2.1. Time-span

The period under survey ranges from 1700 to 1900. The dates coincide with both scientific and historical important events. From a historical point of view, the dates coincide with two important events, the restoration of the English monarchy and the end of the Victorian era. Concerning science and with a direct effect on scientific discourse, events such as the foundation of the Royal Society or the application of the experimental method using also induction and mathematics to explain scientific experiments can be mentioned. The way of writing scientific topics evolved when the old method of speculating and reaching conclusions through introspection gave way to one of objective empiricism, verifying hypotheses through controlled experiments. Francis Bacon was the main proponent of this new method and Robert Boyle played an important role in its evolution (Gotti 1996, 2001; Hunter 2007; Lareo/Montoya 2007). Fulton (1932: 78) highlighted the contribution of the scientist, describing Boyle's undertaking "to establish Science as an integral part of the intellectual life of ordinary men". He encouraged scientists to write in a language that their contemporaries could understand. Moreover, outlined in his Proëmial Essay (1661) the features experimental essays should have. Following his guidelines most of the authors included in CETA express in their prefaces the way of communicating scientific ideas in "so very plain and so comprehensive a manner" (Watts 1726: viii), "ease of expression" (Watts 1726: xiii), "in which the most useful and interesting parts of the science are clearly and familiarly explained" (Bonnycastle 1786: title page), "in as popular and easy a manner as the nature of the subject would admit (Bonnycastle 1786: iv) or "in a pleasant easy and familiar way" (Harris 1719: title page).

Both centuries have been divided into small sub-periods spreading over ten years. Thus each century contains ten decades comprising 20 sub-periods altogether. Each sub-period includes two different texts and authors, as is explained in the following section.

2.2. Texts and authors

CETA includes two texts on Astronomy per decade originally written in English by English native-speaking authors. We have tried to collect samples that could be considered representative of all possible forms (genres or text types) published at that time (McEnery/Wilson 1996; Biber *et al.* 1998: 251-253). Books containing either a collection of letters or of short articles were common at that time: letters were then considered as a more or less informal means of communication. However, as science and its way of expression changed, letters also changed their style and gave way to a more formal vehicle of communication in modern times known as journal article. Bearing these changes in mind we have included essays, textbooks, lectures, manuals, letters, dialogues, treatises, articles, encyclopaedias and other minor vehicles of communication more difficult to classify.

The selection of a wide variety of genres or text types allows the researchers to classify their results taking into account a more or less specialised discourse. Although the samples themselves are not classified as such, we have included the genre or text type they represent to offer researchers a wider range of possibilities. The text types included in CETA could be divided into two blocks according to their specialization (table 2).

+ SPECIALISED	*- SPECIALISED*
Treatises	Dialogues
Articles	Lectures
Essays	Letters
Other (dictionary)	Textbooks

Table 2. Specialisation within genres/text types in CETA.

Only first editions, when available, were selected for the corpus because they better represent the language of the sub-period in which they were written. In some cases, other editions were accepted but only when their dates of publication were not more than 30 years later, period in which Kytö/Rudanko/Smittenberg (2000: 92) assume a language change can be observed.

The option of including only one text per author was preferred, in order to increase the representativeness of idiolectal variants. In this sense, some of the compilation principles of the *Lampeter Corpus of Early Modern English Tracts* were followed. Although our decisions to exclude translations, to include only one work by author and to use first editions made the compilation work more difficult, they strengthened the representativeness and accuracy of CETA to the period under survey.

Concerning authors, we have tried to include well-known as well as "unkown" scientists representing both genders. The fact that authors were famous or not famous had a direct consequence on the information about their lives. That is, in the cases in which the author is unknown, we could not find any information about their lives. In those few cases of the eighteenth century we have assumed that the author belongs to the country where the book was published. The representation of both genders in CETA was also a difficult task. Astronomy or the observation of the sky and stars was not considered a decent task; therefore, it was not easy to find female astronomers signing their works. Besides, female authorship, if any, was not always public at that time. Some exceptions are the English Margaret Bryan and the Irish Agnes Mary Clerke, both included in CETA. Unfortunately, the manuscripts of the American Maria Mitchell[2] and Elizabeth M. Bardwell or the British Mary Ashley were not available at that time. Another important astronomer was the German Caroline Lucretia Herschel, sister of the famous William Herschel. She was the first woman that could publish some of her writings in the *Philosophical Transactions of the Royal Society*, but despite her important discoveries, she was never invited to become a member. It was not until 1945 that the Royal Society changed this situation (Schiebinger/Condor 2004: 377). Fortunately, these prejudices do not affect all the disciplines included in the CC. In Life Sciences, for instance, it was easier for us to find works written by women, perhaps because it includes some of the tasks that could be considered "more feminine", for instance, the observation and description of plants.

Another important detail about authors was their place of education. Although the authors should belong to North America (NA) or Great Britain (GB), we considered that the most important information about their language is where they have acquired their linguistic habits. Therefore, it is more accurate for us to inform about where they went to school or to university instead of their place of birth, which is less relevant. Table 3 lists authors included in CETA, the year of publication and the place of education.

---

2 The first woman to win an international medal, the first woman elected to the American Academy of Arts and Sciences, the first woman member of the American Association for the Advancement of Science, and the first woman professor of Astronomy (Gregory Kohlstedt 1978: 39).

AUTHOR	*YEAR*	*PLACE OF EDUCATION*
Henry Curson	1702	(GB)
Robert Morden	1702	(England) (GB)
William Whiston	1715	Cambridge (England) (GB)
John Harris	1719	Oxford (England) (GB)
George Gordon	1726	(GB)
Isaac Watts	1726	Southampton/Stoke Newington (England) (GB)
Samuel Fuller	1732	(Ireland) (GB)
Jasper Charlton	1735	(GB)
Roger Long	1742	Cambridge (England) (GB)
James Hodgson	1749	(GB)
John Hill	1754	Westminster, (England) (GB)
James Ferguson	1756	Banffshire (Scotland) (GB)
Matthew Stewart	1761	Glasgow/Edinburgh (Scotland) (GB)
George Costard	1767	Oxford (England) (GB)
Alexander Wilson	1774	St. Andrews (Scotland) (GB)
George Adams	1777	London (England) (GB)
John Lacy	1779	(GB)
William Nicholson	1782	North Yorkshire (England) (GB)
John Bonnycastle	1789	London (England) (GB)
Samuel Vince	1790	Cambridge (England) (GB)
Margaret Bryan	1797	London (England) (GB)
Robert Small	1804	Dundee (Scotland) (GB)
John Ewing	1809	New London/New Jersey (Penssylvania /New Jersey (NA)
David Brewster	1811	Jedburgh/Edingburgh (Scotland) (GB)
William Phillips	1818	London (England) (GB)
John Gummere	1822	Horsham/Westtown (Pennsylvania) (NA)
Thomas Luby	1828	Dublin (Ireland) (GB)
John Frederick William Herschel	1833	Windsor/Cambridge (England) (GB)
Landon Cabell Garland	1838	Lexington (Virginia) (NA)
Denison Olmsted	1841	Farmington/New Haven (Connecticut) (NA)

AUTHOR	*YEAR*	*PLACE OF EDUCATION*
Duncan Bradford	1845	(NA)
William Holms Chambers Bartlett	1855	West Point (New York) (NA)
William Whewell	1858	Lancaster/Cambridge (England) (GB)
Ormsby McKnight Mitchel	1860	Ohio (Ohio) (NA)
Elias Loomis	1868	New Haven (Connecticut) (NA)
William Chauvenet	1871	Philadelphia/New Haven (Pennsylvania/ Connecticut) (NA)
Dorman Steele	1874	Syracuse (New York) (NA)
George Howard Darwin	1880	Cambridge (England) (GB)
Charles Augustus Young	1880	Hanover (New Hampshire) (NA)
James Croll	1889	Perthshire (Scotland) (GB)
Agnes Mary Cerke	1893	Cork/Dublin (Ireland) (GB)
Percival Lowell	1895	Cambridge (Massachusetts) (NA)

Table 3. Place of education.

Graph 1 shows the geographical distribution, while the Appendix includes bibliographic details of the samples.

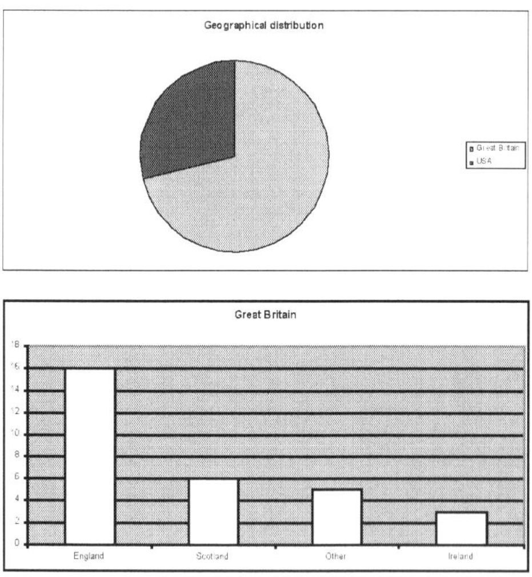

Graph 1. Geographical distribution.

2.3. Samples

Another important decision when compiling the corpus was the length of the samples included. Although Biber (1993: 243-257) affirms that 1,000 words could be enough to study most of the linguistic questions, we have decided to include around 10,000 words per sample to encompass even those topics that could not be so common in scientific discourse. Thus, when it is completed, CETA will contain 409,909 words: 208,079 for the eighteenth century and 201,830 for the nineteenth century (graph 2).

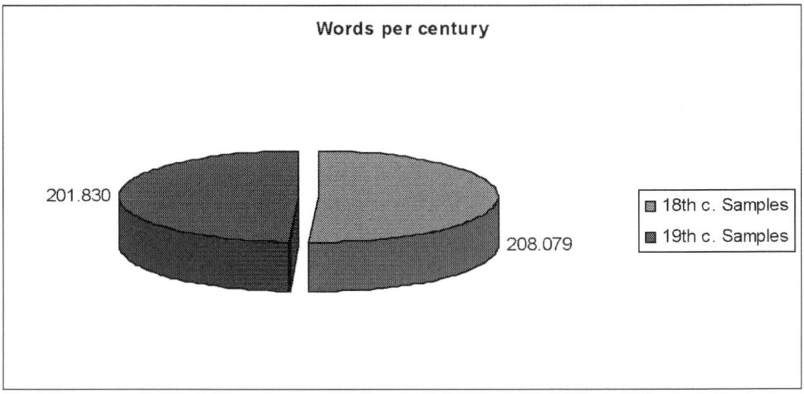

Graph 2. Words per century.

The last step in the compilation of the corpus was the conversion of the original texts into electronic files. Therefore, the next task undertaken was the keying of the texts using XML *eXtended Markup Language* to codify them. All the tags used were developed by the *Text Encoding Initiative* (TEI) and are available for TEIlite (graph 3). We have used the Unicode Standard to represent most symbols and old characters (graph 4); that allows *CETA* to offer a closer representation of the original text.

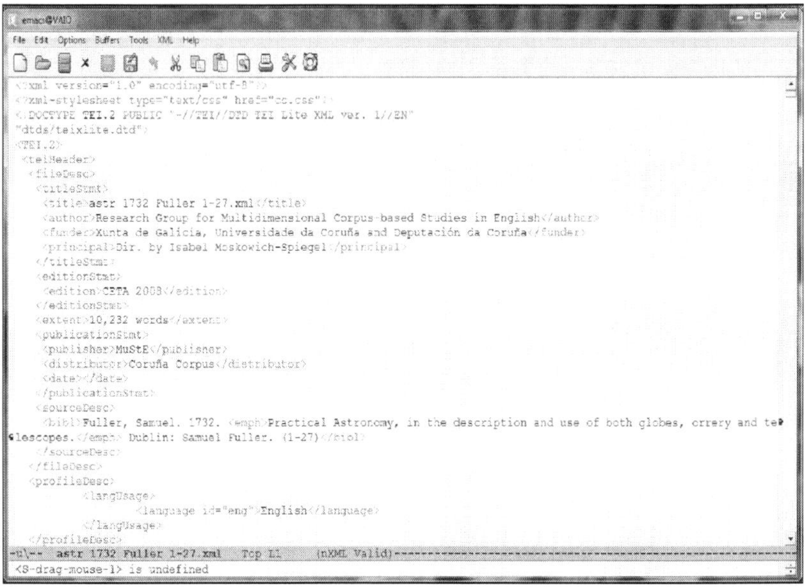

Graph 3 . Text encoded and tagged with TEI in xml format.

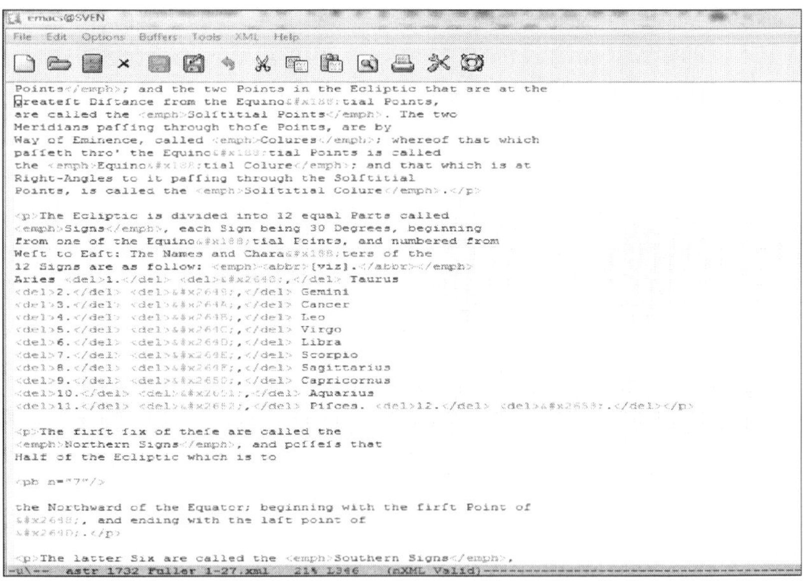

Graph 4. Xml file with Unicode symbols.

Unicode symbols allow us to represent all the graphic variants used across the centuries such as long <s>, <ct> ligature, <æ>, <œ>, etc. When searching one of these symbols, the CCT will retrieve all the possibilities found in the corpus. Graph 5 displays the search for "phenomena" that retrieves the different variants written by the authors (with <æ>, <e> and <œ>).

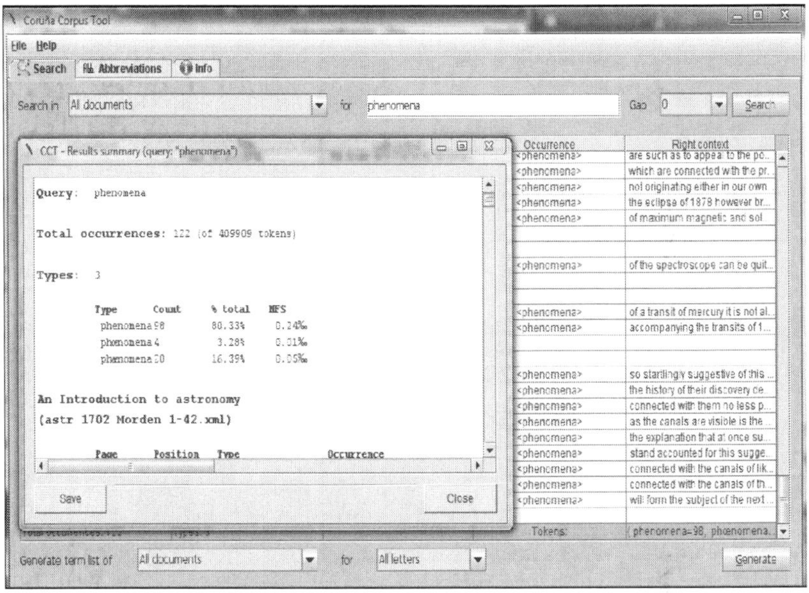

Graph 5. Retrieval of all spelling variants.

3. The *Coruña Corpus Tool* (CCT)

The CCT is a tool that allows the researcher to explore and to work with the CC. It was developed by the *Information Retrieval Lab* (IrLab) in collaboration with the MuStE group of the University of A Coruña (Parapar/Moskowich 2007; Moskowich/Parapar 2008; Lareo 2009b). To give more accurate results, the CCT will index the texts using only the analysable elements of the corpus. The use of wildcards and the inclusion of "metadata", linguistic and extralinguistic information about authors and texts, give the researcher access to valuable information to get more accurate results.

3.1. Editorial decisions taken

Our first idea was for the electronic texts to show the original samples selected truthfully. However, the decision of devoting each text to the language used only by the author himself and to avoid the inclusion of a great number of unnecessary elements compelled us to make some changes. Consequently, we have deleted editors' additions such as footers, margin notes, or headers. We have also deleted quotations, poems, and symbols or numbers that do not have a syntactic function. Although the deleted material is not included into the final count of words, it is shown in one of the CCT's windows.

To exploit all the advantages of the CCT and to make the researchers' work much easier, we have made some layout changes. For instance, in the case of early works, we have deleted unnecessary blank spaces before punctuation marks. We have also corrected obvious spelling mistakes after checking in the OED all the variants that could have been used at that time. Due to xml restrictions, footnotes will appear below the word they refer to instead of at the bottom of the page. The most important edition is the addition of editorial marks to identify the place where something was deleted or to make the work of the CCT more efficient. These editorial marks can be found in the following cases:

-[quotation], [poem], [table], [fragment], [figure], etc. are used to identify the place where deleted parts appear in the original text, such as quotations, tables, poems, figures, fragments; or
-[note] and [endnote] identify the beginning and end of footnotes.

All editorial marks are written between square brackets and some of them can be seen in different CCT's windows. But the square brackets are also used to identify parts of the texts or to disambiguate. Thus, when abbreviations appear in the text we enclosed them into square brackets. They are also elements of disambiguation to distinguish between lexical elements and formulae or numbers that could form an English word. For instance, the roman number I, which uses the same character as the first person pronoun "I", must be written [I], or examples such as the sentence: *the points* [CAB] *form a triangle*. Where *CAB* could be taken as the word *cab* by the CCT. After working with the CCT and checking the results, we hope that the editorial decisions taken will make it possible to combine successfully historical research and computer tools.

3.2. A trip through the CCTool working with CETA

The CCT has three windows, which offer different possibilities (graph 6): Number 1 is the search window, where searches can be made using different parameters. By clicking on the bottom row of this window, number 4, a summary window appears, showing all the results and percentages found in the corpus. In number 2, the abbreviations window, all the abbreviations used in the texts are displayed; and number 3 is the info window, where information about the criteria followed to compile CETA, the metadata about authors and texts as well as a closer version to the original texts are displayed.

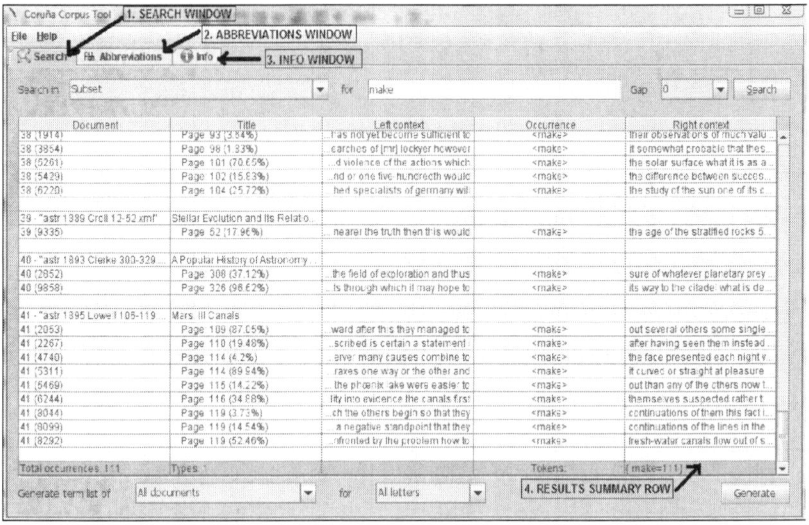

Graph 6. First look at the CCT.

The search window (number 1) shows also information about the title of the text, name of the author, the page where the example appears, the occurrence and its left and right context, and the number of occurrences and different variants found (on the bottom row, number 4). On describing this window, it is also important to note that clicking on the example row a window with the context appears highlighting the word searched (graph 7).

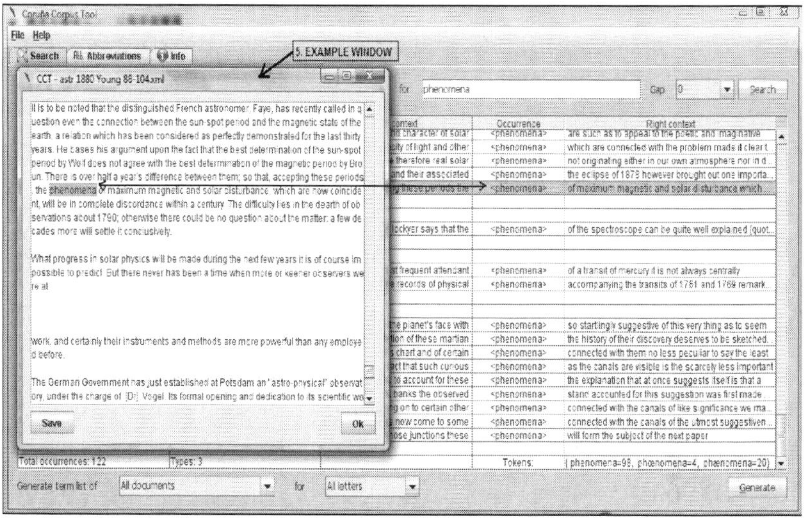

Graph 7. Example window.

The CCT also offers the possibility of working with all the documents included in CETA, only with one of them or with a group of selected texts. To work with a group of texts, you have to choose "subset" on the "search in" block. After selecting it, the options offered will appear (graph 8, right side). It is possible to make a personal selection of texts clicking on the corresponding texts, but the CCT includes also a wide range of parameters to work with texts that follow the same criteria. Clicking on the "metadata bottom", the parameters appear (graph 8, left side). The parameters can be applied to the authors or to the texts. Working with authors, the selection can be made taking into consideration the date of birth and death, the place of education, the gender and the age of the author when the text was published. At the same time, variants such as genre or text type, year of publication and the discipline can be selected. Once the parameters are chosen and applied, the metadata window must be closed to work with the texts that fulfil those requirements.

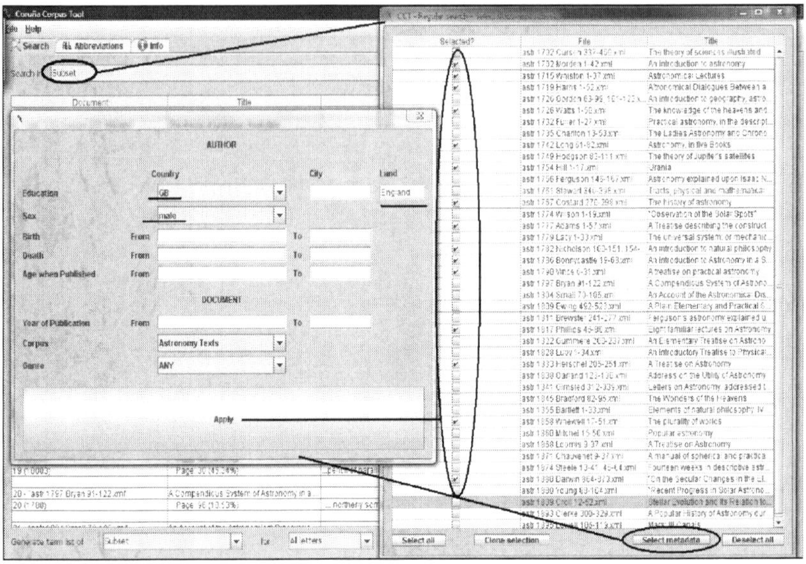

Graph 8. Subset window applying metadata.

In the info display window (graph 9) the metadata and a version closer to the original texts are available. Here it is possible to have a look at biographic details about the authors (graph 9) and extra information about the samples (graph 10). An edited version of the texts is also shown (graph 11).

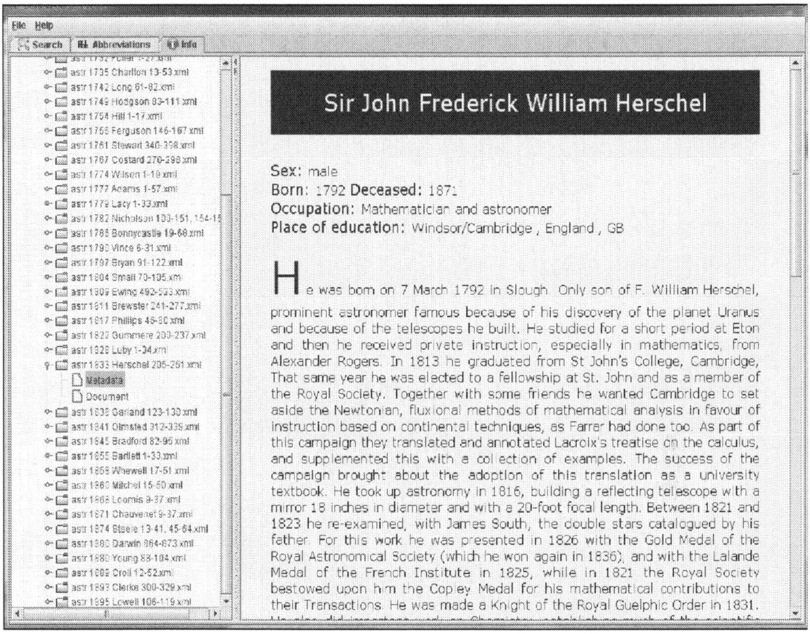

Graph 9. Metadata about the author.

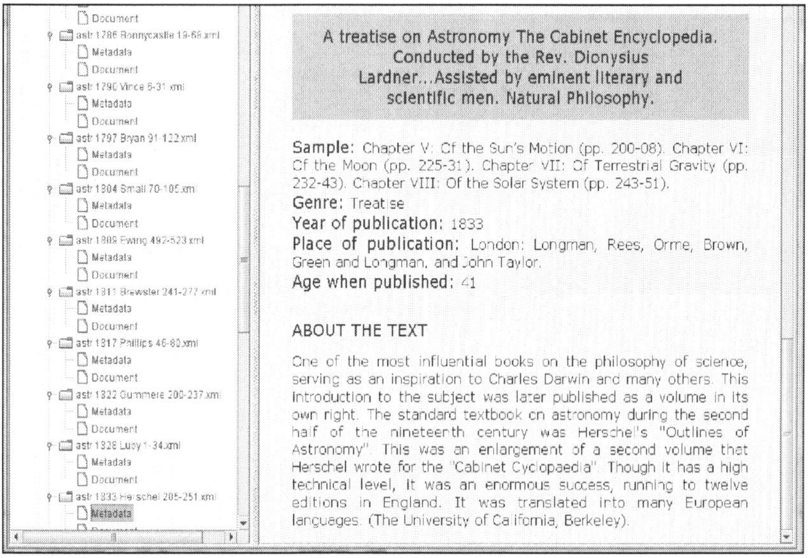

Graph 10. Metadata about the text.

Graphs 11 and 12 show the display of the document on the Info window. Clicking on the label "document" it is possible to see a version closer to the original text with the deleted parts such as paragraph numbers, quotations (in red) and the editorial marks for parts of the text that were written in Greek ([fragmentgreek]) or for footnotes, such as [note], [endnote], or the square brackets used to identify abbreviations (Arab. Græc.).

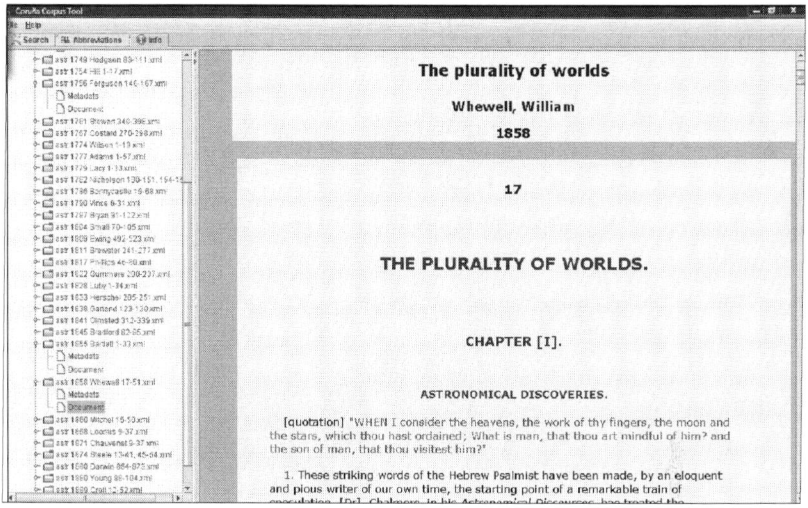

Graph 11. Display of the document.

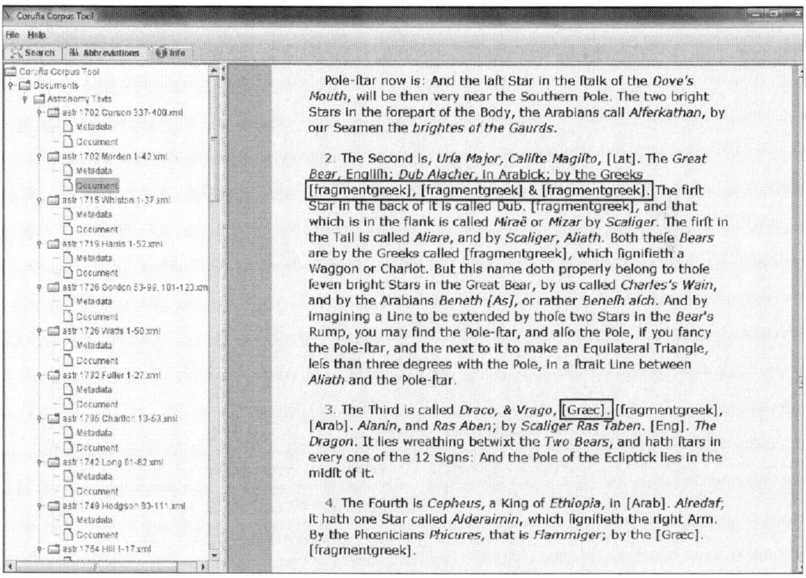

Graph 12. Display of the document with editorial marks.

CETA is part of a bigger Project, the *Coruña Corpus of English Scientific Writing*, aiming at offering a general view of the development of scientific English in several different disciplines from the beginning of the eighteenth century onwards. We hope that the decisions taken when compiling the CETA and the tool developed for its deep exploration will help the linguistic community to make a comprehensive study of the language of Astronomy from a historical point of view.

References

Biber, Douglas 1993. Representativeness in Corpus Design. *Literary and Linguistic Computing*. 8/4, 243-257.

Biber, Douglas / Conrad, Susan / Reppen, Randi 1998. *Corpus Linguistics*. Cambridge: Cambridge University Press.

Bonnycastle, John 1786. *An Introduction to Astronomy in a Series of Letters from a Preceptor to His Pupil, in Which the Most Useful and Interesting Parts of the Science Are Clearly and Familiarly Explained.* London: printed for J. Johnson.

Fulton, John F. 1932. Robert Boyle and His Influence on Thought in the Seventeenth Century. *Isis*. 18, 77-102.

Gotti, Maurizio 1996. *Robert Boyle and the Language of Science*. Milano: Guerini Scientifica.

Gotti, Maurizio 2001. The Experimental Essay in Early Modern English. *European Journal of English Studies*. 5/2, 221-139.

Gregory Kohlstedt, Susan 1978. The Advancement of Women in Science. *New England Quarterly*. 51/1, 39-63.

Harris, John 1719. *Astronomical Dialogues between a Gentleman and a Lady: Wherein the Doctrine of the Sphere, Uses of the Globes, and the Elements of Astronomy.* London: T. Wood.

Hunter, Michael 2007. Robert Boyle and the Early Society: A Reciprocal Exchange in the Making of Baconian Science. *British Journal for the History of Science (BJHS)*. 40/ 1, 1-23.

Kytö, Merja / Rudanko, Juhani / Smitterberg, Erik 2000. Building a Bridge between the Present and the Past: A Corpus of 19-century English. *ICAME*. 24, 85-97.

Lareo, Inés 2008. Analysing a Type of Collocation. *Make* Complex Predicates in 19th Century Science and Fiction. *Revista Canaria de Estudios Ingleses*. 57, 165-180.

Lareo, Inés 2009a. *Make*-collocations in Nineteenth-Century Scientific English. *Studia Neophilologica*. 81, 1-18.

Lareo, Inés 2009b. El *Coruña Corpus*. Proceso de compilación y utilidades del *Corpus of English Texts on Astronomy* (CETA). Resultados preliminares sobre el uso de Predicados complejos en CETA. In Cantos Gómez, Pascual / Sánchez Pérez, Aquilino (eds) *A Survey on Corpus-based Research*. Murcia: Asociación Española de Lingüística de Corpus, 267-280.

Lareo, Inés / Esteve Ramos, Mª José 2008. 18th Century Scientific Writing: A Study of *Make* Complex Predicates in the *Coruña Corpus*. *ICAME Journal*. 32, 69-96.

Lareo, Inés / Montoya Reyes, Ana 2007. Scientific Writing: Following Robert Boyle's Principles in Experimental Essays – 1704 and 1998. *Revista Alicantina de Estudios Ingleses*. *RAEI*. 20, 119-137.

McEnery, Tony / Wilson, Andrew 1996. *Corpus Linguistics*. Edinburgh: Edinburgh University Press.

Moskowich, Isabel / Crespo García, Begoña 2007. Presenting the *Coruña Corpus*: A Collection of Samples for the Historical Study of English Scientific Writing. In Pérez Guerra, Javier *et al.* (eds) *'Of Varying Language and Opposing Creed': New Insights into Late Modern English*. Linguistic Insights Series. Bern: Peter Lang, 341-357.

Moskowich, Isabel / Parapar López, Javier 2008. Writing Science, Compiling Science. The *Coruña Corpus of English Scientific Writing*. In Lorenzo Modia, María José (ed.) *Proceedings from the 31st AEDEAN Conference*. A Coruña: Universidade da Coruña, 531- 544.

Parapar, Javier / Moskowich, Isabel 2007. The Coruña Corpus Tool. *Procesamiento del Lenguaje Natural*. 39, 289-290.

Schiebinger, Londa / Condor, María 2004. *¿Tiene sexo la mente? Las mujeres en los orígenes de la ciencia moderna.* Valencia: Universitat de Valencia.

Watts, Isaac 1726. *The Knowledge of the Heavens and the Earth Made Easy: Or, the First Principles of Astronomy and Geography Explain'd by the Use of Globes and Maps: With a Solution of the Common Problems by a Plain Scale and Compasses as well as by the Globe.* London: printed for J. Clark and R. Hett.

Appendix

Adams, George 1777. *A Treatise Describing the Construction and Explaining the Use of Celestial and Terrestrial Globes.* London: printed for the author.

Bartlett, William Holms Chambers 1855. *Elements of Natural Philosophy (Spherical Astronomy).* New York: Barnes & Co.

Bonnycastle, John 1786. *An Introduction to Astronomy in a Series of Letters from a Preceptor to his Pupil, in Which the Most Useful and Interesting Parts of the Science Are Clearly and Familiarly Explained.* London: printed for J. Johnson.

Bradford, Duncan 1845. *The Wonders of the Heavens, Being a Popular View of Astronomy, Including a Full Illustration of the Mechanism of the Heavens; Embracing the Sun, Moon, and Stars.* Boston: American Stationers Co.

Brewster, Sir David 1811. *Ferguson's Astronomy Explained upon Sir Isaac Newton's Principles: with Notes and Supplementary Chapters.* Philadelphia: Printed by and for A. Small.

Bryan, Margaret 1797. *A Compendious System of Astronomy in a Course of Familiar Lectures: In Which the Principles of the Science Are Clearly Elucidated so as to Be Intelligible to Those Who Have Not Studied Mathematics: Also Trigonometrical and Celestial Problems, with a Key to the Ephemeris, and a Vocabulary of the Terms of Science Used in the Lectures Which Later Are Explained Agreeably to Their Application in Them.* London: printed for the author.

Charlton, Jasper 1735. *The Ladies Astronomy and Chronology in Four Parts*. London: Thomas Gardner.

Chauvenet, William 1871. *A Manual of Spherical and Practical Astronomy, Embracing the General Problems of Spherical Astronomy, the Special Applications to Nautical Astronomy, and the Theory and Use of Fixed and Portable Astronomical Instruments, with an Appendix on the Method of Least Squares*. Vol I. Philadelphia: J. B. Lippincott & Co.

Clerke, Agnes Mary 1893. *A Popular History of Astronomy in the Nineteenh Century*. London: Adam & Charles Black.

Costard, George 1767. *The History of Astronomy, with its Application to Geography, History, and Chronology; Occasionally Exemplified by the Globes*. London: James Lister.

Croll, James 1889. *Stellar Evolution and its Relation to Geological Time*. New York: Appleton.

Curson, Henry 1702. *The Theory of Sciences Illustrated, or the Grounds and Principles of the Seven Arts; Grammar, Logick, rhetorick, Musick, Arithmetick, Geometry, Astronomy. Accurately Demonstrated and Reduced to Practice*. With a variety of questions, problems and propositions both delightful and profitable. London: Richard Smith.

Darwin, George Howard 1880. On the Secular Changes in the Elements of the Orbit of a Satellite, Revolving about a Tidally Distorted Planet. *Philosophical Transactions*. London.

Ewing, John 1809. *A Plain, Elementary and Practical System of Natural Experimental Philosophy: Including Astronomy and Chronology*. Philadelphia: Hopkins and Earle.

Ferguson, James 1756. *Astronomy Explained upon Sir Isaac Newton's Principles and Made Easy to Those Who Have Not Studied Mathematics*. London: printed for the author.

Fuller, Samuel 1732. *Practical Astronomy, in the Description and Use of Both Globes, Orrery and Telescopes wherein the Most Useful Elements, and Most Valuable Modern Discoveries of the True Astronomy Are Exhibited, After a Very Easy and Expeditious Manner, in an Exact Account of our Solar System, with Ten Curious Copper-plates*. Dublin: printed for the author.

Garland, Landon C. 1838. An Address on the Utility of Astronomy. *Southern Literary Messenger; Devoted to Every Department of Literature and the Fine Arts*. Richmond, VA: T.W. White (etc.), 4/2.

Gordon, George 1726. *An Introduction to Geography, Astronomy, and Dialling. Containing the Most Useful Elements of the Said Sciences, Adapted to the Meanest Capacity, by the Description and Uses of the Terrestrial and Celestial Globes with an Introduction to Chronology.* London: J. Senex; G. Strahan; W. and J. Innys; J. Osborn and T. Longman.

Gummere, John 1822. *An Elementary Treatise on Astronomy. In Two Parts. The First, Containing a Clear and Compendious View of the Theory. The Second, a Number of Practical Problems.To Which Are Added, Solar, Lunar and Some Other Astronomical Tables.* Philadelphia: Kimber and Sharpless.

Harris, John 1719. *Astronomical Dialogues between a Gentleman and a Lady: Wherein the Doctrine of the Sphere, Uses of the Globes, and the Elements of Astronomy.* London: T. Wood.

Herschel, John F. W. 1833. A treatise on Astronomy. *The Cabinet Encyclopedia. Conducted by the Rev. Dionysius Lardner ... Assisted by Eminent Literary and Scientific Men. Natural Philosophy.* London: Longman, Rees, Orme, Brown, Green and Longman, and John Taylor

Hill, John 1754. *Urania: or, a Compleat View of the Heavens; Containing the Antient and Modern Astronomy, in Form of a Dictionary: Illustrated with a Great Number of Figures Comprising all the Constellations, with the Stars Laid Down According to Their Exact Situations and Magnitudes: From Repeated and Accurate Observations (Vol. I. Being the First of a Compleat System of Natural and Philosophical Knowledge*). London: T. Gardner.

Hodgson, James 1749. *The Theory of Jupiter's Satellites: With the Construction and Use of the Tables for Computing their Eclipses.* London: W. and J. Mount and T. Page on Tower-Hill, and H. Whitridge.

Lacy, John 1779. *The Universal System: Or Mechanical Cause of all the Appearances and Movements of the Visible Heavens: Shewing the True Powers Which Move the Earth and Planets in Their Central and Annual Rotations with A Dissertation on Comets, the Nature,*

Cause, Matter, and Use of their Tails, and the Reasons of Their Long Trajectories; Likewise and Attempt to Prove What it is that Moves the Sun around its Axis. London: J. Auckland.

Long, Roger 1742. *Astronomy, in Five Books*. Cambridge: printed for the author.

Loomis, Elias 1868. *A Treatise on Astronomy*. New York: Harper and Brothers.

Lowell, Percival 1895. Mars: Canals. *The Atlantic Monthly: Mars III. Canals*. 76/7, 106-109.

Luby, Thomas 1828. *Introductory Treatise on Physical Astronomy*. London/Dublin: Baldwin and Cradock.

Mitchel, Ormsby McKnight 1860. *Popular Astronomy. A Concise Elementary Treatise on the Sun, Planets, Satellites and Comets*. New York: Phinney, Blakeman & Mason.

Morden, Robert 1702. *An Introduction to Astronomy, Geography, Navigation, and other Mathematical Sciences Made Easier by the Description and Uses of the Celestial and Terrestrial Globes. In Seven Parts*. London: Printed for R. Morden, at the Atlas in Cornhil.

Nicholson, William 1782. *An Introduction to Natural Philosophy. Illustrated with Copper Plates*. London: Printed for J. Johnson

Olmsted, Denison 1841. *Letters on Astronomy, Addressed to a Lady in Which the Elements of the Science Are Familiarly Explained in Connexion with its Literary History. With Numerous Engravings*. Boston: Marsh, Capen, Lyon and Webb.

Phillips, William 1818. *Eight Familiar Lectures on Astronomy: Intended as an Introduction to the Science: For the Use of Young Persons and Others Not Conversant with the Mathematics*. New York: James Eastburn and Co.

Small, Robert 1804. *An Account of the Astronomical Discoveries of Kepler: Including an Historical Review of the Systems Which Had Successively Prevailed before his Time*. London: J. Mawman.

Steele, Joel Dorman 1874. *Fourteen Weeks in Descriptive Astronomy*. New York: A.S. Barnes.

Stewart, Matthew 1761. *Tracts, Physical and Mathematical. Containing an Explanation of Several Important Points in Physical Astronomy; And a New Method for Ascertaining the Sun's Distance from the Earth, by the Theory of Gravity.* Edinburgh: printed for W- Sands and A. & J. Bell. London: A. Millar and J. Nourse.

Vince, Samuel 1790. *A treatise on Practical Astronomy.* Cambridge, Mass: J. Archdeacon, J. and J. Merrill, J. Nicholson and W. Lunn.

Watts, Isaac 1726. *The Knowledge of the Heavens and the Earth Made Easy: Or, the First Principles of Astronomy and Geography Explain'd by the Use of Globes and Maps: With a Solution of the Common Problems by a Plain Scale and Compasses as well as by the Globe.* London: printed for J. Clark and R. Hett.

Whewell, William 1858. *The Plurality of Worlds. With an Introduction by Edward Hitchcock. I- Astronomical Discoveries - p17/ II Astro-nomical Objection to Religion.* New York/Boston: Sheldon, Blakeman & co.

Whiston, William 1715. *Astronomical Lectures, Read in the Publick Schools at Cambridge.* London: R. Senex and W. Taylor.

Wilson, Alexander 1774. Observations on the Solar Spots. *Philosophical Transaction.* 64, 1-30.

Young, Prof., LL. D., Ph. D. 1880. Recent Progress in Solar Astronomy. *The Princeton Review.* 1, 88-104.

Gonzalo Camina

New Words for New Ideas: Noun Formation in the *Corpus of English Texts on Astronomy*[1]

1. Introduction

In recent times English has become the lingua franca of science. Scientific terms coined in English have been adopted by other prestigious languages, and new terms are created by means of different devices that make use of native elements of the English language. But the status of English as "a universal language" has not always been the same. Only three centuries ago, English had Latin, Greek and French as its main donors regarding technical terms. Hogben describes the vernacular at the end of the eighteenth century as "a medium of communication [that] had still at least a smattering of a schoolboy Latin" (1969: 11). The reason for this change needs to be searched for in the history of the language and science. In the sixteenth century a progressive key movement took place in England, which culminated in the seventeenth century in what some historians of science call the Scientific Revolution (Butterfield 1957; Henry 1996). Natural philosophers gradually abandoned Latin in favour of the vernacular, and a new method of conveying ideas and discoveries to a wider audience was put forward in the form of the experimental essay (Gotti 2001).

1 The research here reported on has been funded by Xunta de Galicia: (a) Red de grupos de investigación "Lingua e literature inglesa e identidade" (Consellería de Educación e Coordenación Universitaria, 2007/000145-0), (b) "Recompilación e etiquetaxe electrónica de textos científico-técnicos en lingua inglesa entre os séculos XVII e XX (2): Coruña Corpus", Programa de promoción xeral de investigación do Plan galego de investigación, desenvolvemento e innovación tecnolóxica (Incite), PGIDIT07PXIB104160PR; and by Gobierno de España: "Etiquetado electrónico de textos científico-técnicos en lengua inglesa entre los siglos XVII y XX (2) Coruña Corpus" (FFI2008-01649/FILO), Ministerio de Ciencia e Innovación. These grants are hereby gratefully acknowledged.

In the seventeenth century the so-called Scientific Revolution established the theoretical and methodological foundations of modern science in England. Advancements in mechanics provided a solid ground for the development of other sciences such as astronomy and, as a chain-reaction consequence of more powerful lenses that allowed astronomers to pierce deeper into the skies, it also brought about the necessity to reformulate the new acquired knowledge, which led to a further development of natural philosophy. Natural philosophers adopted the use of the scientific method, a more mathematical analysis of nature, and embraced the idea of developing a precise language that could explain the new realities with a specialised level in the vernacular. In 1664 this atmosphere led the Royal Society to commission the bishop and philosopher John Wilkins to develop a comprehensive philosophical language (Hogben 1969; Subbiondo 2001).

New tools, experiments, discoveries and ideas required a more descriptive language. As a direct consequence, new nouns that could denominate these new realities were highly needed. Several resources were exploited to achieve this goal: semantic shifts to recycle items already existing in English, loans from foreign languages –especially from Latin and Greek–, and word-formation processes, such as compounding and affixation.

The aim of this paper is to analyse the processes employed in the coining of new nouns in Early Modern English scientific writing, paying special attention to affixation as a means to enlarge the vocabulary inventory. The criteria adopted for this purpose stays in line with the quantitative analysis used by Culpeper/Clapham (1996: 199-218), which involves a series of computer searches using the online version of the Oxford English Dictionary (1989). The paper is organised in four main sections. Section 2 presents the materials employed for our study corpus, twenty samples belonging to the eighteenth-century block of the *Corpus of English Philosophical Texts* (CEPhiT), a subset of the *Coruña Corpus of English Scientific Writing*. The analysis of the data is presented in Section 3 and the concluding remarks are explained in Section 4.

2. The corpus

The twenty samples belong to the eighteenth century. Therefore, they are fairly representative of the consolidation of language change during the whole Early Modern English period. Each sample in the corpus contains roughly 10,000 words. The total number of words is 200,902, and the details can be seen in Table 1.

Sample	*Words*	*Nouns*	*%*
1700	10,076	1,727	17.1
1705	10,060	1,889	18.8
1710	10,059	1,852	18.4
1717	10,071	1,929	19.2
1727	10,122	1,885	18.6
1730	10,045	1,631	16.2
1733	10,040	1,904	19
1736	10,049	1,895	18.9
1740	10,030	2,134	21.3
1748	10,019	2,196	21.9
1754	10,023	1,830	18.3
1755	10,031	2,289	22.8
1764	10,032	2,002	20
1769	10,064	2,705	26.9
1774	10,017	2,158	21.5
1776	10,007	2,088	20.9
1780	10,059	2,324	23.9
1790	9,993	2,490	24.9
1792	10,058	2,239	22.3
1793	10,047	2,083	20.7
Total	200,902	41,250	20.5

Table 1. Corpus samples and size.

The twenty documents in question are Mary Astell's *Some Reflections upon Marriage* (1700), George Cheyne's *Philosophical Principles of Natural Religion* (1705), John Dunton's *Athenianism* (1710), Anthony Collins' *A Philosophical Inquiry Concerning Human Liberty* (1717), Robert Greene's *The Principles of the Philosophy of the Expansive and Contractive Forces* (1727), Robert Kirkpatrick's *The Golden Rule of Divine Philosophy* (1730), John Balguy's *The Law of Truth* (1733), Joseph Butler's *The Analogy of Religion* (1736), George Turnbull's *The Principles of Moral Philosophy* (1740), David Hume's *Philosophical Essays Concerning Human Understanding* (1748), Henry Bolingbroke's *The Philosophical Works of the late Right Honorable Henry St. John* (1754), Francis Hutcheson's *A System of Moral Philosophy* (1755), Thomas Reid's *An Inquiry into the Human Mind* (1764), Adam Ferguson's *Institutes of Moral Philosophy* (1769), Edmund Burke's *Thoughts on the Cause of the Present Discontent* (1770), George Campbell's *The Philosophy of Rhetoric* (1776), Catharine Macaulay's *A Treatise on the Immutability of Moral Truth* (1783), William Smellie's *The Philosophy of Natural History* (1790), Mary Wollstonecraft's *Vindication of the Rights of Women* (1792), and Alexander Crombie's *An Essay on Philosophical Necessity* (1793).

All these texts were written originally in English by major philosophers of their day. They belong to the same scientific category of natural philosophy, and cover a varied range of topics, such as religion, morals, love, the mind, and natural history, among others. The authors, both male and female, were born and educated in different English-speaking countries, and wrote in English, though we can attribute to them a knowledge of the classical languages as well. The miscellany of topics covered by the samples and the variety of authors, backgrounds and genders are relevant to this study in order to obtain a broader discursive scope. Each decade is represented by two samples, so that we can trace the diachronic evolution of processes in the eighteenth century. Besides, a variety of genres will be provided, all of which will contribute to offer a wider range of possibilities in the language choice.

The samples contain 200,902 words, out of which 41,250 noun tokens have been extracted. 3,831 are different noun types, which produce 41,250 noun tokens, approximately 20.5 percent of the whole subject matter of the corpus.

3. Analysis of data

On a first level, all nouns have been extracted from the main body of the corpus, with a total of 41,250 tokens and 3,831 types. These types have been further analysed in order to identify complex nouns (Adams 1973, 2001; Bauer 1983). This gives us the result of 1,208 non-derived and 2,623 derived nouns, considering complex all those nouns that have undergone a process of word formation, namely, affixation[2], zero derivation, reanalysis and compounding. It is our intention to give a detailed picture of the evolution of these processes across the Early Modern English period (eModE), though references to Old English (OE) and Middle English (ME) will be offered in terms of comparison. Noun types have been grouped depending on the year of their first citation in the OED, in case they have been registered at all. This can be seen in Table 2.

	Affix.	%	*Zero*	%	*Rean.*	%	*Comp*	%
OE	86	69.3	23	18.5	1	0.8	14	11.3
ME	1,249	87	117	8.2	12	0.8	53	3.7
eModE	815	76.2	150	14	7	0.7	96	9
Total	2,150	82	290	11.1	20	0.8	163	6.1

Table 2. Complex words in CEPhiT (eighteenth century).

The data related to affixation in Table 2 seem somewhat paradoxical. In spite of the fact that the scientific revolution is regarded as a very productive source of new linguistic items, the coining of new nouns by means of affixation falls by almost 11 percent in eModE in our corpus. According to our figures, the deriving of nouns reached its peak in the 1400s and has ever since dropped its productivity gradually till the end of the period covered by this research. The reasons for this decrease at the end of the eighteenth century are not self-evident, though it may be related to a combination of the rise of other processes, and also the need

2 The process known as neoclassical compound (Adams 1973; Bauer 1983), or the combination of initial and final combining forms, will be included in the derivational analysis.

for 18th-century noun consolidation in subsequent centuries. Some of the samples dating from the beginning of the century cannot benefit from the words coined after they were published, so they still make use of older words. This is nonetheless an obscure issue that may need some further study, and will be dealt with when we compare CEPhiT with other scientific disciplines of the *Coruña Corpus*. The evolution of noun affixation in ME and eModE can be observed in Graph 1.

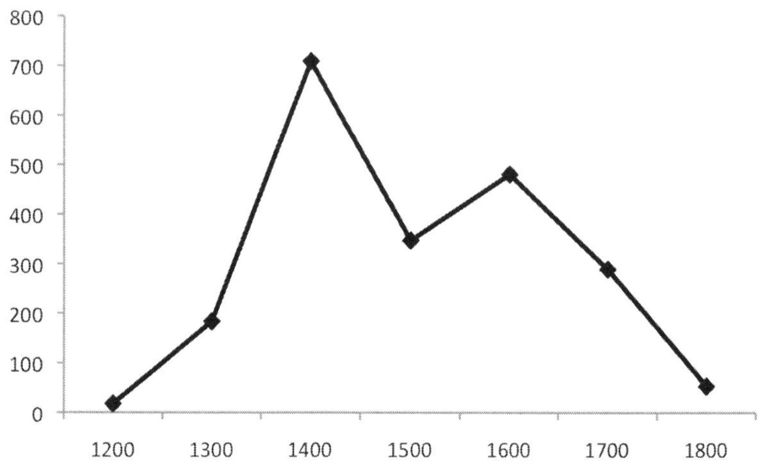

Graph 1. Affixation.

Zero derivation had been a healthy process in OE, but suffered a considerable descent when it came in contact with French. In eModE it recovered to become a steady word-formation process again. This process does not involve the presence of an affix, so it will not be taken further in this paper. Reanalysis in its various forms of back-derivation, clipping or shortening (Adams 2001) did not have the magnitude it has today, though some examples are present in the corpus on their own, such as *chymist* (< *alchemist*), *fog* (< *foggy*), or compounded with other elements, as in *chickweed* (< *chicken* + *weed*).

Compounding poses an interesting line of investigation, since its productivity increases considerably during the Early Modern English period. It might be considered as one of the first indicators of conscious

"Englishness" in word formation for specific purposes, and has grown exponentially until our days, being now one of the main sources of scientific words, especially in the fields of astronomy and particle physics (Raad 1989). The evolution of compounding in ME and eModE can be seen in Graph 2.

At a second stage, derived nouns have been decomposed into their basic elements, and then classified according to their origin and coinage dates.

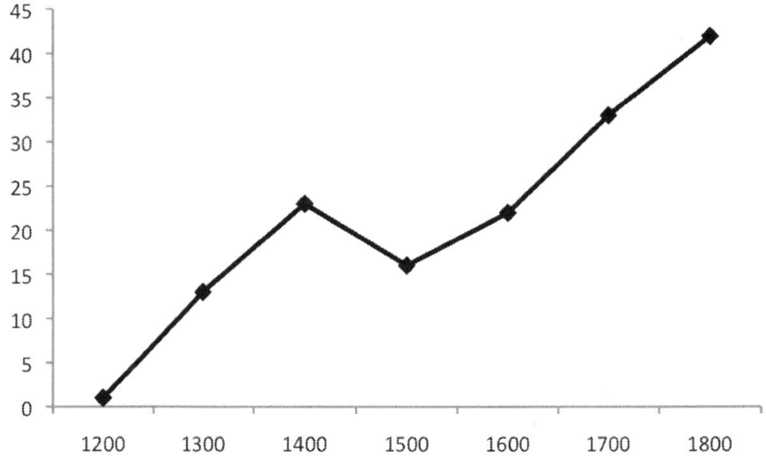

Graph 2. Compounding in ME and eModE.

From this point onwards, we will focus our scope on derivational processes involving affixes only. Nevalainen has pointed out the difficulty of determining what can be considered a derived word and what should be regarded as a loan in Early Modern English (2006: 52), mainly because not only bases but also affixes were adopted from romance languages. Consequently there is not a precise boundary between a loan from French, for instance, and a base imported from French, to which a French affix is attached. It is not our intention to delimit these imprecise borders between loan and affixation for this study, therefore all noun types liable to be decomposed into base+affix or affix+base elements will be taken as derived; however, we have distinguished between three types of derivation for our records: derivation 1 comprises all those complex nouns whose bases had been coined and used in English prior to the derived form; derivation 2 includes complex nouns whose bases had *not* been recorded in the OED before their derived form; and derivation 3 deals with the already mentioned prefix+suffix combination that some authors have labelled neoclassical compounding.

This subdivision may seem confusing at first, but it can shed some light on the decrease of affixation in EModE mentioned earlier. The figures in our analysis show that, in spite of the fact that more new nouns are derived in ME than in eModE, most of these nouns are classified under our derivation 2 group, as can be seen in Table 3.

	Types	*%*	*Tokens*	*%*
ME	678	25.8	10,966	26.6
eModE	204	7.8	1,415	3.43

Table 3. Derivation 2.

The figures shown above illustrate several factors: first, the higher number of ME derivation 2 noun types is obvious to a certain extent, because it is closer in time to the language from which the bases were adopted, and there are fewer written records that can attest their first use in English as bases. The substantial difference in tokens can be attached to the consolidation of words that had already become part of the vocabulary of English. It is worth noting here that this can also imply that nouns coined in eModE might be more specialised and, therefore, less likely to be used frequently. However, we must bear in mind that ME comprises 400 years in the history of the English language, and some of the samples presented for this study may only attest for the years 1500–1700, because they were written near the turn of the century, that is, at the beginning of the eighteenth century.

In the following paragraphs, we will give an account of the category and origin of the bases involved in the formation of nouns in the corpus, and the origin and productivity of prefixes and suffixes. Tables 4 and 5 show the bases and the word categories in ME and eModE in the corpus. OE bases have been omitted in Table 4 because they are originally OE themselves.

Language	*ME types*	*%*	*eModE types*	*%*
French	630	51	327	41.1
Latin	172	13.9	298	37.5
Greek	1	0.1	5	0.6
Italian	0	0	7	0.9
AF/AN	252	20.4	37	4.7
Romance	1,055	85.4	674	84.8
Scandinavian	1	0.1	1	0.1
Swedish	-	-	1	0.1
Norse	-	-	4	0.5
Dutch	2	0.2	-	-
German	-	-	1	0.1
OE	99	8	24	3
ME	69	5.6	31	3.9
eModE	-	-	49	6.2
Germanic	171	13.8	111	14
Turkish	-	-	1	0.1
uncertain	9	0.7	9	1.1
Total	1,235		795	

Table 4. Base origin.

Language	*OE*	*%*	*ME*	*%*	*eModE*	*%*
Adjective	37	51.4	114	21	168	27.8
Noun	10	13.9	84	15.5	137	22.7
Proper Noun	-	-	6	1.1	14	2.3
Verb	24	33.3	327	60.2	266	44
Ppl. Stem	-	-	5	0.9	9	1.5
Ppl. Adjective	1	1.4	7	1.2	10	1.7
Total	72		543		604	

Table 5. Base category.

As regards base origin, both Romance and Germanic components have followed a uniform trend since the twelfth century (Table 4). The romance component is present in around 85 percent of the types, whereas the Germanic component is relegated to around 14 percent. French is the main donor of bases in both periods, responsible for almost half the types. Latin increases its presence in eModE, taking over the Anglo-Norman influence of the beginning of the Middle English period.

Base category in Table 5 demonstrates that verbs are the most productive as regards affixation in the two later periods, with a peak of 60 percent in ME, whereas adjectives lead derivations in nouns coined in the OE period. It is worth mentioning here, though, that bases that were not coined before their derivative counterparts have not been taken into consideration. This means that 678 types in ME and 204 in eModE[3] have been left out of this study, and may have altered these results significantly.

We will conclude our analysis of the samples by providing tables with the prefixes and suffixes[4] involved in derivative processes in our corpus, together with their origin and degree of productivity. They can be seen in Tables 6 and 7.

Prefix	*Origin*	*OE*	*ME*	*eModE*
anti-	G	-	-	2
apo-	G	-	-	1
arch(i)-	L	1	-	2
astro-	G	-	2	-
bio-	G	-	-	1
co-	L	-	-	1
counter-	ME	-	-	3
de-	L	-	1	1
demi-	F	-	-	1
dia-	G	-	2	1
dis-	L	-	5	17
eco-	G	-	1	-
enter-	F	-	-	1
equi-	L	-	1	1

3 Included in the "derivation 2" group in our database.

4 We have respected the classification of affixes in the OED.

Prefix	*Origin*	*OE*	*ME*	*eModE*
fore-	OE	1	3	1
geo-	G	-	2	1
gymno-	G	-	1	-
hyper-	G	-	-	2
hypo-	G	-	2	2
im-(1)	L	-	-	2
im-(2)	L	-	1	-
in-(2)	L	-	1	10
in-(3)	L	-	7	2
inter-	L	-	-	1
ir-(2)	L	-	-	1
matri	L	-	1	-
meta	G	-	-	1
micro	G	-	-	1
mis-(1)	OE	-	1	6
mis-(2)	F	-	1	-
non-	AN	-	-	5
off-	OE	-	-	1
omni-	L	-	-	1
onto-	G	-	-	2
out-	OE	-	3	2
over-	OE	-	-	1
para-	G	-	1	4
patri-	L	-	1	-
peri-	G	-	1	-
poly-	G	-	-	1
pre-	L	-	1	2
re-	L	-	4	6
semi-	L	-	-	1
soli-	L	-	-	1
sub-	L	-	2	1
super-	L	-	1	2

Prefix	*Origin*	*OE*	*ME*	*eModE*
sur-	F	-	2	2
sym-	G	-	-	2
syn-(2)	G	-	-	1
theo-	G	-	-	1
thermo-	G	-	-	1
tri-	L	-	1	-
un-(1)	OE	-	3	2
uni-	L	-	1	-
zoo-	G	-	-	1
Total		2 prefixes 2 types	30 prefixes 55 types	45 prefixes 101 types

Table 6. Prefixes.

In our corpus this type of derivation involves 77 different prefixes and initial combining forms (Bauer 1983), which produce 158 noun types. Greek and Latin provide the majority of them, with 21 each; French, 4; Anglo-Norman (Anglo-French), 2; Old English, 6; and Middle English, 1. There is a constant increase not only in the number of prefixes involved in affixation, but also in the number of types generated by these prefixes. In contrast to what happens to some OE suffixes, which have stopped being productive altogether, prefixes have survived throughout the three periods studied for this paper.

Suffix	*Origin*	*OE*	*ME*	*eModE*
-acy	L	-	5	7
-ade	F	-	-	2
-age	F	-	20	6
-al(1)	L	-	8	6
-an	L	-	7	7
-ance	F	-	45	19
-ancy	L	-	1	7
-and	L	-	-	1
-ane	L	-	1	-
-ant	F	-	9	2

Suffix	*Origin*	*OE*	*ME*	*eModE*
-ar(2)	L	-	1	-
-ar(3)	OE	1	1	-
-ard	F	-	6	-
-arian	L	-	-	1
-ary(1)	L	-	-	9
-ate(1)	L	-	11	5
-ate(2)	L	-	1	-
-ation	L	-	49	27
-cide	F	-	2	1
-cism[1]	?	-	-	1
-cosm	G	-	-	1
-c(u)le	F	1	9	6
-dom	OE	2	3	-
-ean	L	-	-	2
-ee(1)	AF	1	1	3
-eer(1)	F	-	-	1
-el(1)	OE	1	1	-
-el(2)	F	-	4	2
-ence	F	-	3	28
-enchyma[2]	G	-	-	1
-ency	L	-	5	22
-ene	?	-	-	1
-ent	F	-	12	4
-er(1)	OE	6	58	35
-er(2)	F	-	21	-
-er(3)	OE	-	1	-
-er(4)	AF	-	-	4
-ery	F	-	9	13
-ess(1)	F	-	6	1
-ess(2)	ME	-	1	-
-et(1)	F	-	7	4
-et(3)	L	-	-	1
-ful	OE	1	2	-

Suffix	*Origin*	*OE*	*ME*	*eModE*
-graphy	G	-	-	2
-hood	OE	1	2	-
-ian	L	-	12	6
-ic	F	-	7	5
-ication[3]	L	-	-	1
-ice(1)	F	-	15	2
-ician	ME	-	1	2
-ics	ME	-	3	9
-ine(1)	L	1	5	4
-ine(4)	F	-	3	1
-ine(5)	F	-	-	1
-ing(1)	OE	13	49	89
-ing(2)	OE	1	-	-
-ing(3)	OE	1	-	-
-ion(1)	L	1	238	112
-ise(2)	F	1	5	-
-ish(1)	OE	1	-	-
-ish(2)	F	-	2	1
-ism	F	-	3	-
-ison	F	-	1	-
-ist	F	-	-	25
-istic	F	-	-	1
-ite(1)	F	-	2	-
-ite(2)	L	-	-	2
-ition	F	-	5	2
-ity	L	-	76	70
-ive	F	-	4	-
-le	ME	1	9	1
-ling	OE	-	-	1
-logue	G	-	1	-
-logy	G	-	2	3
-ment	AN	-	44	39

Suffix	*Origin*	*OE*	*ME*	*eModE*
-meter	F	-	2	1
-metry	F	-	1	1
-mony	L	-	5	1
-ness	OE	22	31	40
-nomy	G	-	2	-
-oon	F	-	1	-
-or	L	1	35	14
-ory	AN	-	4	3
-ose(2)	L?	-	-	1
-ot	F	-	-	1
-our	AN	1	19	1
-pathy	G	-	-	1
-phyte	G	-	-	1
-ry	F	-	3	4
-scope	L	-	-	1
-ship	OE	2	2	4
-sis	G	-	-	1
-sophy[4]	G	-	-	1
-sphere	G	-	-	1
-ster	OE	1	-	-
-th(1)	OE	11	7	2
-tion	F	-	15	5
-tude	L	-	11	2
-ty(1)	ME	-	44	12
-ure	F	-	30	16
-y(1)	OE	4	1	-
-y(2)	OE	-	1	-
-y(3)	F	-	50	21
-y(4)	AF	-	7	1
-y(5)	AF	-	13	-
Total		23 suffixes 76 types	76 suffixes 1,132 types	79 suffixes 715 types

Table 7. Suffixes.

We have recorded 178 different suffixes and final combining forms that produce a total of 1,923 noun types. Latin provides 24; Greek, 11; French, 36; Anglo-Norman (Anglo-French), 7; Old English, 18 suffixes.

Suffixation is not only a source of new words but also a source of new word categories, since it can be both class-changing and class-maintaining. This may be the reason why it is so much more used than prefixation, given that the latter is only class-maintaining, and the derived element belongs to the same category as the non-derived one.

Concerning the periods in question, Early Modern English shows a wider variety in the use of suffixes, which are in turn surpassed by the large amount of suffixation originated in Middle English. Some Old English suffixes have stopped being productive, such as *-dom*, *-ful*, *-hood* or *-th,* the last examples in the corpus being *freedom* (1225), *scornful* (1400), *falsehood* (1297) and *growth* (1557), respectively. Nevertheless, others have survived and produce new derived nouns affixed to both Germanic and Romance bases. This is the case of *-er* (99 types), *-ing* (151) and *-ness* (93). This capacity of Old English affixes to be used in combination with Romance bases is not corresponded by Romance affixes, which do not in general produce new derivatives attached to Germanic bases. One of the few examples that can be found in our corpus, if not the only one (though this has not been studied in depth yet) is *amazement*, coined in 1595.

Regarding the productivity of other suffixes, Latin *-ion* displays the highest number of types, 351; Anglo-Norman *-ment* produces 83 types, and French *-ance*, 64 types. The study of prefixes and suffixes is very interesting, and can provide explanations about the rise and fall of some of them, the alternance of several pairs of coexisting elements, used alternatively by the same author, like *-ence*, *-ency*, *-ance*, *-ancy*; *-or*, *-our*; *-ar*, *-er*, and so forth. Nevertheless, it would be rather extensive to refer to it here, and will be dealt with in future research.

4. Conclusion

This paper has examined the processes of word formation in twenty samples from the 18th century, all of them dealing with the same scientific subject, philosophy, and containing a varied range of topics and authors to provide our study with a variety of linguistic choice. We have tried to overcome the difficulty of analysing derivation processes in a period of the history of the English language where the boundaries between what can be considered as affixation or loan are diffuse, and we have established the patterns of noun formation present in our corpus. These patterns include affixation, zero derivation, reanalysis, compounding, analogy and even mere invention[5]. We have paid attention to the evolution of the derivative processes across three periods, namely Old English, Middle English and Early Modern English. We have attested the decrease in affixation in Early Modern English, together with the increase of other processes such as zero derivation or compounding, which can be interpreted as a first attempt to produce new linguistic items involving elements already existing in the language. Besides, the division between types of derivation used for this analysis confirms that almost half of the derived nouns coined in the Middle English period appeared before the non-derived noun from which they apparently derived. This, for some authors, could be considered as a loan instead of derivation. The same pattern, applied to the Early Modern English period, shows that only approximately one fifth of the derived nouns coined could be considered as loans instead of complex words as a result of affixation.

It is widely acknowledged that, in general, Latin provides English with the highest amount of material for the production of new words in the period studied here (Nybakken 1959; Nevalainen 2006). However, the figures in our database show that, as regards nouns in CEPhiT, French has a larger impact on the philosophical language in the eighteenth century, supplying more affixes than Latin, and also more bases to which these affixes can be attached. It would be interesting to apply the same study to other scientific disciplines to observe whether

5 As is the case of *yahoo*, invented by Daniel Defoe and used by Reid (1764: 28).

this phenomenon is exclusive to philosophy or, on the contrary, can be extrapolated to other sciences.

Finally, Latin affixes produce the highest amount of new derived forms, especially *-ion* and *-ity*, but it is significant to mention the increase in productivity of some Old English suffixes, such as *-er*, *-ness* and, especially, *-ing*, which is by far the most used, and deserves, perhaps, a more detailed study.

References

Adams, Valerie 1973. *An Introduction to English Word-Formation.* London: Longman.

Adams, Valerie 2001. *Complex Words in English*. London: Longman.

Astell, Mary 1700. *Some Reflections upon Marriage*. London: John Nutt, 42-89.

Balguy, John 1733. *The Law of Truth: Or, the Obligations of Reason Essential to All Religion*. London: J. Pemberton, 1-43.

Bauer, Laurie 1983. *English Word-Formation*. Cambridge: Cambridge University Press.

Bolingbroke, Henry 1754. *The Philosophical Works of the Late Right Honorable Henry St. John, Lord Viscount Bolingbroke*. London: D. Mallet, 5-45.

Burke, Edmund 1770. *Thoughts on the Cause of the Present Discontent*. London: J. Dodsley, 1-21.

Butler, Joseph 1736. *The Analogy of Religion, Natural and Revealed, to the Constitution and Course of Nature*. London: Knapton, 1-32.

Butterfield, Herbert 1957. *The Origins of Modern Science*. New York: The Free Press.

Campbell, George 1776. *The Philosophy of Rhetoric*. London: W. Strahan and T. Cadell, 1-55. Vol I. Book I.

Cheyne, George 1705. *Philosophical Principles of Natural Religion: Containing the Elements of Natural Philosophy, and the Proofs for Natural Religion, Arising from Them*. London: G. Strahan, 1-58.

Collins, Anthony 1717. *A Philosophical Inquiry Concerning Human Liberty*. London: R. Robinson, 45-115.

Crombie, Alexander 1793. *An Essay on Philosophical Necessity*. London: J. Johnson, 1-66.

Culpeper, Jonathan / Clapham, Phoebe 1996. The Borrowing of Classical and Romance Words into English: A Study Based on the Electronic Oxford English Dictionary. *International Journal of Corpus Linguistics*, vol. 1, nº 2, 199-218.

Dunton, John 1710. *Athenianism: Or, the New Projects of Mr. John Dunton*. London: T. Darrack, 1-25, 330-334.

Ferguson, Adam 1769. *Institutes of Moral Philosophy*. Edinburgh: Printed for A. Kincaid, J. Bell, 50-154.

Gotti, Maurizio 2001. The Experimental Essay in Early Modern English. *European Journal of English Studies*. 5/2, 221-239.

Greene, Robert 1727. *The Principles of the Philosophy of the Expansive and Contractive Forces or an Inquiry into the Principles of the Modern Philosophy, that is, into the Several Chief Rational Sciences, Which Are Extant. In Seven Books*. Book I. Cambridge: Cambridge University Press, 1-13

Henry, John 1997. *The Scientific Revolution and the Origins of Modern Science*. London: MacMillan.

Hogben, Lancelot 1969. *The Vocabulary of Science*. London: Heinemann.

Hume, David 1748. *Philosophical Essays Concerning Human Understanding*. London: A. Millar, 1-50.

Hutcheson, Francis 1755. *A System of Moral Philosophy*. Vol. II. Book III. Glasgow: R. and A. Foulis, 149-185.

Kirkpatrick, Robert 1730. *The Golden Rule of Divine Philosophy*. London: R. Kirkpatrick, 7-37.

Macaulay Graham, Catharine 1783. *A Treatise on the Immutability of Moral Truth*. London: A. Hamilton, 1-54.

Nevalainen, Terttu 2006. *An Introduction to Early Modern English*. Oxford: Oxford University Press.

Nybakken, Oscar 1959. *Greek and Latin in Scientific Terminology*. Iowa: Iowa State University Press.

Raad, B.L. 1989. Modern Trends in Scientific Terminology: Morphology and Metaphor. *American Speech*, vol. 64, no. 2, 128-136.

Reid, Thomas 1764. *An Inquiry into the Human Mind, On the Principles of Common Sense*. Edinburgh: A. Millar, 1-60.

Smellie, William 1790. *The Philosophy of Natural History*. Edinburgh: Heirs of C. Elliot, 1-54.

Subbiondo, Joseph L. 2001. Educational Reform in Seventeenth-Century England and John Wilkins' Philosophical Language. *Language & Communication*. 21, 273-284.

The Oxford English Dictionary (Second Edition). 1989. Oxford: Oxford University Press.

Turnbull, George 1740. *The Principles of Moral Philosophy. An Enquiry into the Wise and Good Government of the Moral World*. London: John Noon, 1-33.

Wollstonecraft, Mary 1792. *Vindication of the Rights of Woman: With Structures on Political and Moral Subjects*. London: J. Johnson, 15-68.

Estefanía Sánchez Barreiro

Adjunctive and Disjunctive Lists in Modern English Scientific Discourse[1]

1. Introduction

From early Modern English up to these days, an important number of enumerations end with constructions such as *or such like, and so on, &c.* This paper presents a first approach to the study of these structures classified under the name of extenders. There is a widespread opinion that general extenders make no contribution to communication. Nevertheless, this does not seem to be the case since general extenders have a major impact on what is communicated (Overstreet 1999: 21).

According to Ruth Carroll (2007), extenders are phrases which consist of a conjunction and a non-specific noun phrase. In general, this type of constructions is identified by the use of conjunctions like *and* or *or*, although they are not exclusive (*with*). In this paper, we focus on phrases collocating with *and* and *or*.

Our aim is to observe the most frequent extenders scientific authors used in the eighteenth century in order to create an initial inventory. To that end, first we decided to build up a list with the possible tokens which helped us delimit our search. Next, we chose ten eighteenth century scientific texts taken from the *Coruña Corpus of English Scientific Writing*.[2] All of them pertain to the same scientific discipline: Life Sciences.

---

1 The research reported on here was funded by Universidade da Coruña and the Xunta of Galicia through its Consellería de Innovación e Industria, Dirección Xeral de Investigación e Desenvolvemento, grant number PGIDIT07PXIB104160PR, supervised by Isabel Moskowich.

2 However, we must note that, since the *Coruña Corpus* is an ongoing project it is possible that some data may be altered in the future.

2. Definition of extenders

Although extenders have received little attention by linguistic researchers, they are in fact relevant as a discourse strategy since, contrary to what could be expected, users would resort to them in multiple circumstances. As Overstreet (1999: 3) confirms, extenders serve a function which varies according to contexts of use. They typically occur at clause-final position and their basic form corresponds to conjunction plus noun phrase. Overstreet (1999) chose the name of general extenders to make reference to such expressions, but the truth is that there are preferable options for other authors who have used terms like *extension particles*, *set-marking tags* (Dines 1980; Ward/Birner 1993), *list completers* (Jefferson 1990; Lerner 1994) and, even, *vague category identifiers* (Channell 1994).

In this first approach, we have preferred the term 'extenders' to define these particular expressions. All of them correspond to phrases like *and stuff like that, or the like, and other, and so on, and so forth* which help, in a way, "extend" meaning.

Conjunctions like *and* or *or* are typically but not exclusively found in extenders. We may also find structures beginning with the preposition *with*, though, in the analysis, we will keep to phrases introduced only by *and* and *or* in initial position.

Although previous studies (Overstreet 1999; Carroll 2007) have confirmed that the frequency of occurrence of extenders is higher the more informal the context is, we also chose to examine these structures in the written scientific medium.

3. Adjunctive and disjunctive constructions

Overstreet (1999: 3) establishes a basic division of extenders: those beginning with *and* called 'adjunctive' which would include *and that, and so on* and those beginning with *or* called 'disjunctive' including *or what, or something*. We will analyse these two types of constructions.

The action of having made a distinction between general and specific extenders was due to the fact that we thought it to be essential since as we carried out our research we saw there would be hardly a few occurrences in case of only taking general extenders, which finally repeated over and over again in all our samples. This way, by including specific extenders, our possibilities increased.

Extenders are not exclusive of English, but they are frequent in many different languages. In fact, extenders may be found in different contexts which go from a newspaper article or the lyrics of a song to a message in an answering machine or radio news, both in spoken and written contexts, more formal or more informal (although they are more likely to occur generally in informal, spoken English). That is the reason why this sort of studies has been primarily based on spoken samples. For instance, in the case of Cheshire (2007) her investigations were based on data taken from adolescent speech.

On the contrary, Carroll (2007) used a corpus of letters (*Corpus of Early English Correspondence*) as the main source for her findings.[3] In fact this author states that "correspondence is one of the best alternatives" if we lack in comparable data.

4. General and specific extenders

It is difficult to establish distinctions between general extenders and specific extenders. It seems clear, for instance, that constructions like *or anything of that kind*, *or anywhere* or *and stuff like that* are good examples of general extenders.

We can distinguish general extenders when the phrase used refers to additional members but in an unspecified way. That is to say, the phrase is more general and covers a wider range of possibilities as regards meaning. General extenders, in this case beginning with *and* or *or*, can include nouns

3 Görlach (1999: 149) also considers that making linguistic studies through letters is particularly interesting in this case as these ones reflect, to a great extent, social and functional relations between sender and addressee and this, he says, is only something comparable to oral texts.

like *stuff* or *things* after the conjunction, or even indefinite pronouns and adverbs. The following example will show this more clearly:

(1) Some seeds will retain their vegetative property for a considerable time if buried in a bed of earth, a foot or more below the surface (Donovan 1794: 51).

On the other hand, we could classify the phrase *or any other affection of the body* as an example of specific extender. In this case, the phrase delimits the extra possibilities within the whole description, and there is more lexical material expressed.

All in all, one of the clues that must be taken into account when it comes to distinguishing between the two is to make sure whether the extender is accompanied by an adjective or by a relative clause. If this is so, we are undoubtedly dealing with a specific extender. The example below illustrates what we are referring to:

(2) […] or as velvet; it prevents shell-fish or other marine insects from fixing their habitation on those shells (Donovan 1794: 61).

5. Data and methodology

Data for the analysis have been taken from the *Coruña Corpus*, a computer-readable corpus which consists of a collection of samples for the historical study of English scientific writing.

The *Research Group for Multidimensional Corpus-based Studies in English* (MuStE), from the University of A Coruña (Spain), is currently carrying out this ambitious project devoted to the compilation of a corpus of English scientific texts published between 1650 and 1900. The team is in the process of creating a corpus that can be used for the diachronic study of scientific discourse.

The compilation of the *Coruña Corpus*[4] has been, and still is, governed by some of the most common parameters used in Corpus Linguistics, namely, external criteria for the delimitation of dates,

4 For more information see Moskowich/Crespo (2007), and Crespo/Moskowich (forthcoming).

sampling techniques, number of words per sample, etc. It covers as well different disciplines including the one we will use for the present study: Life Sciences.

Extenders are hidden expressions as they may exist in a text but are not directly identifiable. To sort the problem out we decided to create an inventory of possible forms consulting previous works based either on spoken data (Overstreet 1999) or on written data (Carroll 2007).

This paper can be interpreted as a challenge in the study of extenders since it is based on data taken from academic formal writing. However, texts of such level of technicality could considerably reduce the number of extenders found, we do not expect to find a significant quantity, but at least we hope the results obtained are relevant enough as to shed light on which ones were most frequently used at that time and why.

As explained before, our samples will cover a number of texts between 1707 and 1794 pertaining to the subject matter of Life Sciences. Table 1 below shows that the number of words per text is around 10,000 words, hence our corpus will be constituted by 100,428 words.

	AUTHOR	DATE OF PUBLICATION	NUMBER OF WORDS
TEXTS EXTRACTED FROM THE *CORUÑA CORPUS* FOR THIS STUDY	Douglas, James	1707	10,025
	Keill, James	1717	10,014
	Blair, Patrick	1723	10,051
	Boreman, Thomas	1730	10,002
	Blackwell, Elizabeth	1737	10,077
	Dodd, James	1752	10,045
	Bancroft, Edward	1769	10,073
	Withering, William	1776	10,101
	Speechly, William	1786	10,017
	Donovan, Edward	1794	10,023
TOTAL OF WORDS			100,428

Table 1. Corpus material.

As was stated earlier, we will search for general and specific extenders collocating with the conjunctions *and* and *or* within a particular and closed list. In order to do that, we prepared that list with the most common extenders other authors described in their respective works. Such is the case of Overstreet (1999: 4) and Aijmer (2002: 221). It is worth mentioning we have always taken into account the historical period of our texts and, although based on what others found, we carefully chose all of the possible tokens due to our contextual possibilities.

Fortunately, we could count on the *Coruña Corpus Tool*[5], which simplified our search and made it easier. Using this software, we could quickly identify all the tokens suggested in the lists. The next step consisted in discarding all those misinterpreted occurrences manually as not every token was valid for our study.

Apart from that, it is worth saying that we have not taken into account any other general extender out of the list, apart from the cases in which any adverb or other kind of modifier might appear between conjunction and NP.

We first checked that all the expressions were perfectly acceptable in the eighteenth century as they really were. However, we also started from the premise that scientific texts tend to be formal, so we thought it unlikely, if not impossible, to find tokens involving expressions like *and stuff like that* or *and whatnot*, as texts of this kind are usually more formal linguistically.

6. Analysing samples: results

According to Barber (1993: 199) it was from approximately 1700 that language began to behave as we know it today, so it shares many

5 The *Coruña Corpus Tool* is the name given to the software which is going to be used for information retrieval in our texts. The *Coruña Corpus Tool* (CCT) has been developed in collaboration with the *Information Retrieval Laboratory* (IRLab) at the University of A Coruña. It is still being created and has been developed as an aid for linguists so that they can extract and condense all the necessary information about the corpus. For more information see Parapar & Moskowich (2007).

similarities with present-day English. As was stated in the introduction, we have used samples dating back to a period which goes from 1707 to 1794, nearly a century in time, which can throw some light on the most significant differences of use as regards these particles. Unfortunately, though not surprisingly, few have been the general extenders found, probably due to the level of formality of this kind of texts. The fact that scientific texts must be extremely precise with their statements does not enable writers to include more informal words or expressions as such feature may ruin the most precise words of prestigious authors and, that way, minimise their credibility. There is the generalised idea that the use of extenders shows uncertainty on the part of speakers, something that surely investigators would try to avoid to protect their reputation.

As regards results, a total of 186 tokens were identified and analysed. The immediate information which can be inferred from the study is that adjunctive lists have proved to be more frequent than disjunctive ones (166 tokens as opposed to 20). Table 2 below shows these results in more detail:

	General extenders with *and*	Specific extenders with *and*	General extenders with *or*	Specific extenders with *or*	Total per text
Douglas, James (1707)	0	1	1	1	3
Keill, James (1717)	2	4	1	0	7
Blair, Patrick (1723)	12	8	2	3	25
Boreman, Thomas (1730)	35	7	1	1	44
Blackwell, Elizabeth (1737)	3	5	0	0	8
Dodd, James (1752)	4	5	0	1	10

Bancroft, Edward (1769)	20	3	0	0	23
Withering, William (1776)	4	7	0	2	13
Speechly, William (1786)	3	3	1	1	8
Donovan, Edward (1794)	32	8	2	3	45
TOTAL	115	51	8	12	186

Table 2. Corpus findings of general and specific extenders in adjunctive and disjunctive lists.

Both authors, Boreman (1730) and Donovan (1794), contain a larger number of occurrences (44 and 45 respectively). However, we must take into account that more than 30 of these instances in each case correspond to the adjunctive list *&c.* Therefore, a similar number of tokens can be found in the remaining categories.

The only cases with no tokens at all of either general or specific extenders correspond to the fragments by Bancroft (1769) and Blackwell (1737), the only woman in our subcorpus. In comparison, nevertheless, if we take the total instances per text we notice a considerable distance between both.

As regards the use of extenders, Douglas (1707) is the least prolific of the authors. The writer who opens our list only produced one token per category and no adjunctive general extender. In fact, it could be said that this case does not occur with any of the authors as we find, at least, two examples of general extenders beginning with *and* in each text.

After these brief comments on the first results of the data under survey, much more detailed remarks will be offered in the sections that follow.

6.1. Adjunctive lists

The first results of this study present a list of the general extenders found with an adjunctive construction: *&c.*, *and other things*, *and the rest*. Table 3 contains the inventory of the adjunctive tokens which have been identified:

ADJUNCTIVE LISTS	General extenders	Specific extenders	TOKENS
And (all) the rest	0	3	3
And all (that)	0	14	14
And c. (&c.)	114	0	114
And everything (like that)	0	0	0
And more	0	1	1
And other	0	23	23
And other things	1	0	1
And others	0	9	9
And so	0	0	0
And so forth	0	0	0
And so on	0	0	0
And so on and so forth	0	0	0
And so on and so on	0	0	0
And stuff (like that)	0	0	0
And such or any	0	0	0
And that	0	0	0
And the like	0	0	0
And things like this/that	0	0	0
And things of this/that kind/nature/sort	0	0	0
And this and that	0	0	0
And this/that sort/kind/type of thing	0	0	0
And whatever	0	1	1
And whatnot	0	0	0
TOTAL	115	51	166

Table 3. Adjunctive lists and corresponding findings.

Except for a very few occasions as is the case of *and all* and *and other things* the rest of forms identified were *&c*. What may call our attention here is its widespread use, above all, among some authors. The form *&c*., which stands for the former *etcetera*, appears in its abbreviated form in all samples. Curiously enough, although different spellings of the Latin form can be listed according to the Oxford English Dictionary, we have only found tokens corresponding to the abbreviation *&c*. The forms *et cetera, et caetera* and *etc*. could have been other possibilities for the Latin expression whose meaning could correspond to "and other things" or "and so forth". In modern usages it is also used to mean "and so on" or "and more". In this case the ampersand is a ligature of "et" (Latin for *and*).

The phrase *&c*. was often used to refer to continuation in any series of descriptions. This is, at least, the main function it represents in all our samples. Other extenders like *so forth* or *and so on* could have been good substitutes. However, we found zero tokens of these two phrases whereas *&c*. was highly represented in samples like the ones by Boreman (1730) or Donovan (1794) whose general extenders mostly correspond to this phrase.

(3) The domestick Weesil kill and destroy Rats, Mice, and Moles; as also Hens, Chickens Eggs, &c. (Boreman 1730: 62)

(4) [...] whether taken up by the drag-net, from the sides of rocks, or bottoms of ships, &c. (Donovan 1794: 59)

Except for the cases of *&c*., the frequency of general extenders is less than low as we only found one more token within the samples. It is in the text from 1730 by Thomas Boreman that we find one of the most frequent and recognisable expressions: *and other things*. This is perhaps, the most uncertain and informal expression of all.

(5) [...] striking with her fore Feet, both Dogs and other things, as a Man doth with his Hand. (Boreman 1730: 55)

Specific extenders constitute a more numerous group as regards diversity since we counted 51 tokens distributed among six different occurrences: *and the rest, and all, and other, and more, and others, and whatever*. The most frequent collocations with the conjunction *and* correspond to *other* (23 tokens) and *all* (14). It was common that such phrases were modified by adverbs like *yet* (in the case of *and all*) or adjectives like *many, some, any* and *several* as in examples 7 to 8.

(6) […] and the upper part of the Metacarpal Bones, is furnished with four Muscles; and yet all of them, as Veslingius remarks, terminate in the Bones of the Metacarpus. (Douglas 1707: 109)

(7) […] the Cause of the Elasticity of Bodies, of Fermentations, Dissolutions, Coagulations, and many other of the Operations in Chymistry. (Keill 1717: 101)

(8) […] but not without some Exceptions, as black, white, red, in Ethiopia, and some other Parts of Africa, and Asia. (Boreman 1730: 2)

We have analysed both *and other* and *and others* in the plural as specific extenders and not only as a part or an element of another phrase. In any case, frequency is higher with *and other* and more often than not, both are modified by the adjectives mentioned above.

Lastly, we must make reference to *and more* and *and whatever*, two expressions which appear only once each. This is exemplified by 9 below:

(9) […] that are sold and eaten almost as soon as caught, and whatever some may imagine (Dodd 1752: 60)

It is to be noted that the adjunctive phrase *and whatever* is used as a specific extender and not as a general one. Unfortunately, we had only one token. This means it was not possible to see whether this fact repeated or not.

6.2. Disjunctive lists

We have found 20 instances corresponding with extenders which collocate with the conjunction *or*. The final figure for specific extenders proved to be higher than that for general extenders. Table 4 shows the list with all the tokens found in the samples:

DISJUNCTIVE LISTS	General extenders	Specific extenders	TOKENS
Or anyone/anybody/someone/ somebody/someplace(s)/ somewhere (like that)	0	0	0
Or anything (like that)	0	0	0
Or anything/something of that/this kind/sort	0	0	0
Or more	6	5	11
Or other	2	7	9
Or other things	0	0	0
Or so	0	0	0
Or something (like that)	0	0	0
Or stuff (like that)	0	0	0
Or that	0	0	0
Or the like	0	0	0
Or things (like that)	0	0	0
Or this or that	0	0	0
Or what	0	0	0
Or whatever	0	0	0
TOTAL	8	12	20

Table 4. Disjunctive lists and corresponding findings.

As expected, the proportion of disjunctive lists is considerably lower compared to adjunctive ones. Unfortunately, out of the fifteen possibilities we suggested in this list, only two of them were found: *or more* and *or other*. Finally, it was the phrase collocating with *more* (11 tokens) the one which proved to be the most frequent, followed by the expression with *other* (9 tokens). We found *or more* as a general extender up to six times whereas we registered two instances with *or other*:

(10) […] which passing over the head of the Humerus, grows fleshy, and is so inserted into the inside of that Bone, about an Inch or more below its Neck. (Douglas 1707: 99)

The final figure equals more when it comes to dealing with specific extenders. Regarding these ones, we could identify a total of 12 tokens. Again, the same expressions are the only representatives in the category. As well as in the case of some adjunctive lists, disjunctive ones also occasionally appear modified by some kind of adjective. Nearly fifty per cent of the instances with *or other* correspond to the phrase modified by the adjective *any* as can be seen in examples 10 and 11:

(11) [...] and so used for the ordinary Drink in Scorbutical Cases. Also in the Gout, Gravel, or any other affection of the Body (Blair 1723: 6)

(12) sometimes they are made of cedar wood, but very seldom of deal or any other wood that is soft (Donovan 1794: 53)

On the contrary, we found zero cases of the phrase *or more* attached to any type of modifier. In 13 below we identified a three-part enumeration where this disjunctive expression is only followed by a noun:

(13) [...] here it may continue two, three, or more months (Speechly 1786: 54)

As a last consideration, we should notice that the total number of tokens found in our samples does not necessarily correspond to all the possible extenders since we have limited our research to a particular list. All this implies that further study would be needed to reach more reliable conclusions.

7. Conclusion

General extenders serve a function which varies according to contexts of use. They typically occur in clause-final position and their basic form corresponds to conjunction plus noun phrase, a clause or phrase added to a sentence that amplifies its meaning.

In this study, we have examined the most frequent extenders in ten eighteenth century scientific texts belonging to the discipline of Life Sciences. Regarding the first results, the definitive conclusion we reach is that the most repeated extender is the construction corresponding to an abbreviation: *&c.* Few have been the general extenders found if we leave apart this special case. The abbreviation *&c.* occurs a total of 114 times, whereas the rest of general extenders with the conjunction *and* were only produced once, corresponding to the form *and other things*. This way *&c.* was, by far, the most frequent of adjunctive general extenders in our samples. The rest of examples are collected in the group of specific extenders.

Secondly, adjunctive lists have proved to be more frequent than disjunctive ones. *More* is the most frequent particle together with *or* and *other* as regards the conjunction *and*. Due to the level of formality of this kind of texts, the results obtained are not surprising. Authors had to make an effort to follow a rigorous method, the new scientific method, where both processes and facts must be described clearly and thoroughly. This implied the avoidance of terms which would leave a window open to uncertainty or lack of clarity and definition.

Finally, future research could encompass longer lists or a deeper study of other scientific disciplines like Astronomy or Mathematics. This piece of work is expected to be considered a modest contribution in a field with many unexplored possibilities. Of course, we cannot forget to mention the necessity for further study into lists in this period of history relating to scientific texts as this paper has only made initial steps towards what may be a basic inventory of the most frequent forms at that time.

References

Primary sources

Bancroft, Edward 1769. *An essay on the Natural History of Guiana, in South America*. London: printed by Becket and P. A. Hondt (10,073 words)

Blackwell, Elizabeth 1737. *A Curious Herbal, Containing Five Hundred Cuts of the Most Useful Cuts Which Are Now Used in the Practice of Physick*. London: printed by Samuel Harding (10,077 words)

Blair, Patrick 1723. *Pharmaco-botanologia: Or, an Alphabetical and Classical Dissertation on All the British Indigenous and Garden Plants of the New London.* London: printed by G. Strahan; W. and J. Innys; and W. Mears (10,051 words)

Boreman, Thomas 1730. *A Description of Three Hundred Animals; Viz. Beasts, Birds, Fishes, Serpents, and Insects. With a Particular Account of the Whale-fishery.* London: Printed by Rich Ware; T. Game (10,002 words).

Dodd, James Solas 1752. *An Essay towards a Natural History of the Herring.* London: Printed by T. Vincent. (10,045 words)

Donovan, Edward 1794. *Instructions for Collecting and Preserving Various Subjects of Natural History: As Animals, Birds, Reptiles, Shells, Corals Plants, &c.* London: Printed by the Author (10,023 words).

Douglas, James 1707. *Myographiæ comparatæ specimen: or, a comparative description of all the muscles in a man and in a quadruped.* London: printed by W. B. for G. Strahan (10,025 words)

Keill, James 1717. *Essays on several parts of the animal oeconomy.* London: printed by George Strahan (10,014 words)

Speechly, William 1786. *A treatise on the culture of the pine apple and the management of the hot-house. Together with a description of every species of insect that infest hot-houses, with effectual methods of destroying them.* Dublin: printed by Luke White (10,017 words)

Withering, William 1776. *A botanical arrangement of all the vegetables, naturally growing in Great Britain.* Birmingham: printed by M. Swinney (10,101 words)

Secondary sources

Aijmer, Karin 1995. What Happens at the End of Our Utterances? The Use of Utterance-Final Tags Introduced By *and* and *or*. In Andersen, Henning / Koerner, Konrad (eds) *Papers from the 8th Scandinavian Conference of Linguistics*. Copenhague: Institut for Philologie, Kopenhaven University, 366-389.

Aijmer, Karin 2002. *English Discourse Particles: Evidence from a Corpus*. Amsterdam, The Netherlands: John Benjamins.

Barber, Charles 1993. *The English Language: A Historical Introduction*. Cambridge: CUP.

Brinton, Laurel J. 1996. *Pragmatic Markers in English. Grammaticalization and Discourse Functions*. Berlin: Mouton de Gruyter.

Carroll, Ruth 2007. Lists in Letters: NP-Lists and General Extenders in Early English Correspondence. In Moskowich, Isabel / Crespo, Begoña (eds) *Bells Chiming from the Past,* 37-53.

Channell, Joanna 1994. *Vague Language*. Oxford: Oxford University Press.

Cheshire, Jenny 2005. Syntactic Variation and Beyond: Gender and Social Class Variation in the Use of Discourse-New Markers. *Journal of Sociolinguistics*. 9, 479–508.

Cheshire, Jenny 2007. Discourse Variation, Grammaticalisation and Stuff like That. *Journal of Sociolinguistics*. 11, 155-193.

Crespo, Begoña / Moskowich, Isabel Forthcoming. *CETA* in the Context of the *Coruña Corpus*. *Literary Linguistic Computing*.

Dines, Elizabeth R. 1980. Variation in Discourse – "And Stuff like That". *Language in Society*. 9, 13–33.

Dubois, Sylvie 1992. Extension Particles, etc. *Language Variation and Change*. 4, 179–204.

Fischer, Kerstin 1998. Discourse particles, turn-taking, and the semantics-pragmatics interface. *Revue de Semantique et Pragmatique*. 8, 111–137.

Fischer, Kerstin 2000. *From Cognitive Semantics to Lexical Pragmatics: The Functional Polysemy of Discourse Particles*. Berlin: Mouton de Gruyter.

Görlach, Manfred 1999. *English in Nineteenth-Century England. An Introduction*. Cambridge: Cambridge University Press.

Jefferson, Gail 1990. List-Construction as a Task and Resource. In Psathas, G. (ed.) *Interaction Competence*. Washington, DC: The International Institute for Ethnomethodology and Conversation Analysis and University Press of America, 63-92.

Lareo Martín, Inés / Esteve Ramos, María José 2007. 18th Century Scientific Writing: A Study of *Make* Complex Predicates in the *Coruña Corpus*. *ICAME Journal*. 32, 69-96

Lerner, Gene 1994. Responsive List Construction. A conversational resource for accomplishing multifaceted social action. *Journal of Language and Social Psychology*. 13, 20-33.

Macaulay, Ronald 2005. *Talk that Counts: Age, Gender and Social Class Differences in Discourse*. New York: Oxford University Press.

Moskowich, Isabel / Crespo, Begoña 2007. Presenting the *Coruña Corpus*: A Collection of Samples for the Historical Study of English Scientific Writing. In Pérez-Guerra, J. *et al.* (eds) '*Of Varying Language and Opposing Creed': New Insights into Late Modern English*. Berlin/New York: Peter Lang, 341-357.

Norrby, Catrin / Winter, Joanne 2000. Set Marking Tags – "and stuff". In Henderson, J. (ed.) *Proceedings of the 1999 Conference of the Australian Linguistic Society*. <http://www.arts.uwa.edu.au/LingWWW/als99/proceedings.html>

Norrby, Catrin / Winter, Joanne 2002. Affiliation in adolescents' use of discourse extenders. In Cynthia Allen (ed.) *Proceedings of the 2001 Conference of the Australian Linguistic Society.* <http://linguistics.anu.edu.au/ALS2001/papers/winter_norrby.pdf>

OED on CD-ROM. Version 3.00. 2002. Oxford: Oxford University Press.

Overstreet, MaryAnn / Yule, George 2002. The metapragmatics of "and everything". *Journal of Pragmatics*. 34, 785-794.

Overstreet, MaryAnn 1999. *Whales, Candlelight, and Stuff Like That: General Extenders in English Discourse*. New York: Oxford University Press.

Parapar, Javier / Moskowich, Isabel 2007. The Coruña Corpus Tool. *Revista del Procesamiento de Lenguaje Natural*. 39, 289-290.

Schiffrin, D. 1987. *Discourse markers*. Cambridge: Cambridge University Press.

Stubbe, Maria / Holmes, Janet 1995. 'You know, eh and other "exasperating expressions": an analysis of social and stylistic variation in the use of pragmatic devices in a sample of New Zealand English. *Language & Communication*. 15, 63–88.

Ward, Gregory / Birner, Betty 1993. The semantics and pragmatics of "and everything". *Journal of Pragmatics*. 19, 205-214.

Section II. Translation and Scientific Knowledge

Tomás Conde

Tacit Technique on the Evaluation of Technical Texts

1. Introduction

Over the last few years, translation training has been arranged in many ways. Broadly speaking, texts on various subject matters are believed to require different skills, for they belong to fields that do not coincide with each other as regards terminology, syntax, history and their addressees' level of demand. In some training programs, a slight distinction is made between specialised translation and non-specialised translation or, as it is usually (and wrongly) called, 'general' translation. In other programs, specialised translation is subdivided into broad branches that seem to result from vocational approaches rather than from market demands. These branches consist of legal, official, financial and commercial translation, as opposed to scientific and technical translation – the latter usually includes computer translation and localisation. Irrespective of the suitability of one approach or the other, the specialisation of training involves the specialisation of the corresponding teachers. Do the revision and assessment of students' translations depend on whether these texts are 'general' or technical?

As a rule, texts are distinguished either by their level of difficulty or by their nature. In both cases, research is still scarce, at least in the translation field. In fact, Orozco (2001) laments the little empirical research undertaken on texts' difficulty, for there are studies about the difficulty in reading texts, but not in translating them (Darwish 2001). As a consequence, improving the reliability of exams is still problematic (Campbell/Hale 2003). There are some studies on computer assisted translation or automatic translation, but hardly any on human translation. The subject matter of texts is thought to have an impact on their difficulty, but such an effect has not been proved yet. As a matter of fact, translators must usually face significant problems about some aspects, such as syntactic complexity or references, which are very common in non-specialised texts.

Reiβ (1971) separates content-oriented texts (news, manuals, official documentation) from form-oriented (literary texts) and appellative ones (advertising, electoral pamphlets or satirical prose). Apart from that, two types of texts are usually considered with respect to their nature: pragmatic (or 'technical') and semantic (or 'general') texts. For example, Schiaffino/Zearo (2001), from a professional perspective, distinguish between legal and marketing texts. Riccardi (2002) suggests that technical texts be translated semantically and non-technical texts, communicatively; and, according to Muzii (2006: 17), "different types of documentation need different quality requirements".

As translation itself, translation evaluation might depend on the nature of the texts. Adab (2000) states that pragmatic texts should be assessed according to functionalistic criteria, but also acknowledges that these texts may imply an expressive or affective function. Finally, Klaudy/Károly (2000) recommend that ambitious research be undertaken to compare different types of texts, so that results are more reliable, and truly generalisable conclusions can be drawn.

This paper starts from a wider research about serial translation evaluation. More specifically, it is focused on the differences observed in the behavior and results of several groups while evaluating two types of texts: specialised (technical) and non-specialised ('general').

2. Materials and methods

Four groups of evaluators took part in this project: potential addressees, professional translators, translation teachers and translation students. All the addressees were enrolled in a course on English for Specific Purposes in several Schools of Granada, but there were more science students (30) than social science students (10). The 13 professionals had been working several years as freelance or in-house translators. The teachers worked in several Translation Schools and had studied either in Spanish (5) or Mexican (5) universities. Moreover, the 25 translation students were enrolled in a course on Translatology, and had to carry out the evaluation task as a classroom activity. The evaluation task consisted of 'assessing/correcting/proofreading/editing/revising', according to their beliefs and intuition, four sets of 12 Spanish translations each, corresponding to the four original English texts.

Two originals (DP1 and DP3) dealt with politics, whereas the other two (CT2 and CT4) were technical communications, specifically about industrial procedures for painting. The addressees had to work only on two sets, depending on whether they were social science (DP1 and DP3) or science students (CT2 and CT4). DP1 and DP3 were selected from *Economist.com;* DP1 consisted of several news in brief and DP3 was an article about the European Constitution. On the other hand, the technical texts were selected from an online technical list *(finishing.com)* and had a similar structure: a series of messages between specialists on an internet forum. The translations – written by translation students from the preceding year – had been selected among the poorest to prompt evaluators' reactions. Sets were administered alternating by subject matter (according to the code number: DP1, CT2, DP3 and CT4) so that each set was considered as a single task. Furthermore, translations within each set were randomly ordered and coded. Subjects were allowed to receive and work on the task both in a digital format and on paper.

Together with the evaluation task, a document was enclosed with the following instructions: subjects had to 1) process the translations in the order given; 2) work on a whole set in a single session; and 3) classify the translations into VERY BAD, BAD, GOOD or VERY GOOD. This classification made it possible to contrast the quality of the texts with the subjects' behavior. By converting the quality judgments to figures (very bad, 1; bad, 2; good, 3 and very good, 4), they were used as another variable for the statistical analysis of the evaluations. But there were more to come. First were the times that evaluators 'did something' on the text, in other words, the actions ('any mark introduced by an evaluator on the text'). Then, a distinction was made between PRODUCT-ORIENTED changes (modifications of the body text: inclusions, suppressions and substitutions) and FEEDBACK-ORIENTED changes (information for the translation's author, which included phenomena' classifications, text highlights and linkups or marks used just to introduce comments). In turn, comments were written around the body text; they introduced information that did not become direct part of the text, but contained evaluators' explanations or opinions that were grouped, according to their contribution, as REVISION comments (specific improvements of textual fragments, as alternatives, corrections or solutions) and INVESTIGATION comments (which depended on the specific context of the evaluation task and its instructions, and included communications to the researcher, explanations on the evaluation system and general quality assessments).

Other parameters were connected with the phenomenon: 'what motivated an evaluator to act onto a particular text segment'. On the basis of their nature, phenomena were divided into NORMALISED (there was an authority, normally the Spanish Royal Academy, who sanctioned a proper option; and included spelling, proper nouns, misprints, punctuation, format, concordance, syntax, cohesion, terminology and weights and measures) and NOT NORMALISED (correctness, clarity, usage, perspective, omission and divergent interpretations). Another aspect of phenomena was their saliency, i.e. the biggest or smallest coincidence in the subjects at marking or detecting phenomena. Having taken into account that the most salient phenomena were marked by 40 subjects, the following five categories were considered: SINGULAR or non-coincident phenomena (one evaluator), phenomena of VERY LOW coincidence (two or three evaluators), of LOW coincidence (between 4 and 12 evaluators), of MEDIUM coincidence (between 13 and 25) and of HIGH coincidence (between 26 and 40).

This paper is preceded by previous studies, i.e. Muñoz/Conde (2006) studied the order effects of serial evaluation from several perspectives. The present study also discusses some aspects about the variations in the subjects' behavior between the first and the second set of each subject matter. Further, Conde (2007) analysed the differences in the evaluation of technical and non-specialised texts. But only ten subjects (all translation students) participated in that experiment; therefore, the present study may be useful to support or refute the conclusions suggested then, e.g.: in comparison with non-specialised texts, technical texts received higher quality judgments and contained fewer actions, more feedback-oriented changes and less-coincident phenomena.

All data were entered in a database created ad hoc and analysed later with SPSS (v15.0). Nevertheless, in contrast to the previous research, raw data were multiplied by an index which made it possible to compare the results, for not all subjects had evaluated all texts – for example, the addressees only had to evaluate half of the task – and not all translations were the same length. As a consequence, many of the following data are not absolute values but weighted averages.

3. Results

Once the index had been applied, and the atypical values suppressed, subjects averaged 190.65 actions per text. Table 1 shows the average of actions performed per text in each set, as well as the corresponding averages for each group of evaluators.

	Average	*%*	Addressees	Professionals	Teachers	Students
DP1	*295.27*	*38.72*	246.53	679	332.22	504.66
CT2	*154.96*	*20.32*	260.21	236.96	150.48	225.9
DP3	*171.01*	*22.43*	122.69	159.28	138	240.17
CT4	*141.34*	*18.53*	134.91	149.03	92	169.76

Table 1. Actions.

Subjects started intervening much more often: actions in DP1 accounted for nearly 40% of all. After DP1, percentages were similar, but even so there were more actions in non-specialised sets. When groups were considered separately, the professionals and the teachers reduced the number of actions across sets, whereas the students performed more actions on non-specialised texts and, even though both subgroups reduced the amount of actions across sets, the science addressees intervened more than the social science addressees.

Subjects' changes were either FEEDBACK-ORIENTED (marking the text to introduce a comment, highlighting the text or classifying phenomena) or PRODUCT-ORIENTED (substituting, suppressing or including new text). Table 2 shows that subjects tended to perform a similar number of feedback-oriented changes on the four sets, whereas there were fewer product-oriented changes on technical texts.

	Product				Feedback			
	DP1	CT2	DP3	CT4	DP1	CT2	DP3	CT4
Total average	*177.56*	*53.28*	*80.95*	*53.42*	*117.71*	*97.33*	*90.06*	*87.92*
Addressees	97.47	0	31.53	0	149.07	250.41	91.16	147.54
Professionals	460.21	29.3	0	9.14	218.79	123.74	110.34	93.53
Teachers	35	19.21	15.89	9	181.96	109.41	110.23	73
Students	353.7	103.84	141.06	95.26	150.96	113.02	99.11	79

Table 2. Changes' orientation.

Across sets, all groups reduced the amount of feedback-oriented changes. In contrast, the results regarding product-oriented changes pointed to four different behaviors: the social science addressees reduced the number of these changes between DP1 and DP3, whereas the science addressees did not make use of this type of changes at all; the professionals dramatically reduced their product-oriented changes after DP1, and in DP3 they only performed feedback-oriented changes; the teachers reduced the product-oriented changes across sets and the students performed fewer product-oriented changes on technical sets.

Table 3 shows that REVISION comments (which included alternatives, corrections and solutions) were more frequent in DP1 (average 70.8 per text) than in the other sets (CT2: 37.94; DP3: 43.03 and CT4: 34.94). Nonetheless, there were fewer INVESTIGATION comments (which included vocative, procedural and assessment) precisely in DP1, and they were particularly common in CT2: DP1, 7.5; CT2, 13.41; DP3, 9.17 and CT4, 10.03.

	Revision				Investigation			
	DP1	CT2	DP3	CT4	DP1	CT2	DP3	CT4
Total average	*70.8*	*37.94*	*43.03*	*34.94*	*7.5*	*13.41*	*9.17*	*10.03*
Addressees	25.23	93.20	35.55	18.60	20.64	65.60	28.09	18.47
Professionals	153.13	57.43	21.14	18.09	10.72	16.50	10.28	10.20
Teachers	95.87	53.57	50.13	50.60	9.56	6.25	10.10	8.22
Students	111.83	41.50	40.91	21.91	7.97	7.34	8.31	7.89

Table 3. Comment contribution.

The science addressees introduced more comments – regardless of their type – than the social science addressees. In turn, the teachers and the students reduced the number of revision comments throughout the sets; but they displayed fewer investigation comments (even though these comments were more balanced across sets) on technical sets. With regard to revision comments, the professionals' behavior was close to that of the teachers and the students, but they gathered investigation comments especially in CT2.

Next parameter was the nature of phenomena. Table 4 shows the average number of NORMALISED (for instance, punctuation and format) and NOT NORMALISED (for example, clarity and omission) phenomena per text in every set. On the one hand, the number of normalised phenomena dropped after DP1 and stayed at the same level in the last two sets. On the other hand, the average of not normalised phenomena was similar in technical texts and significantly higher in non-specialised texts, most especially in DP1.

	Normalised				*Not normalised*			
	DP1	CT2	DP3	CT4	DP1	CT2	DP3	CT4
Total average	*150.84*	*66.76*	*98.37*	*66.23*	*136.75*	*70.73*	*62.35*	*64.08*
Addressees	83.71	75.02	28.09	25.05	139.89	117.37	63.07	35.89
Professionals	305.30	98.69	47.09	64.10	354.93	67.50	63.34	39.59
Teachers	138.32	63.17	44.11	32.88	146.00	75.11	82.78	40.13
Students	244.97	100.82	86.57	81.27	251.07	99.09	136.58	84.10

Table 4. Nature of the phenomena.

Across sets, the addressees spotted fewer and fewer phenomena, both normalised and not normalised. The number of phenomena decreased considerably between DP1 and CT2 in the groups involved in the translation sphere, but the decline was even more noticeable within professionals and students than among teachers. The academic groups (teachers and translation students) spotted fewer not normalised phenomena in technical texts, whereas the professionals identified more normalised phenomena in CT4 than in DP3.

As for saliency (Table 5), the most common category in all groups was the phenomena of LOW coincidence (4 to 12 evaluators). In CT4, the gap with the second category (MEDIUM coincidence, 13 to

25 evaluators) was narrower. There were more SINGULAR phenomena in non-specialised texts, and more phenomena of medium and HIGH (26 to 40 evaluators) coincidence in technical sets than in DP3.

Set	Singular	Very Low	Low	Medium	High
DP1	36.95	38.05	115.5	76.9	27.86
CT2	22.37	26.29	53.75	40.07	6.25
DP3	30.64	30.15	72.68	33.74	3.81
CT4	18.67	19.62	48.57	48.26	6.22

Table 5. Saliency per set.

The addressees tended to spot fewer phenomena in the second set that they evaluated than in the first one (Table 6); but, in DP3, the social science addressees marked more SINGULAR phenomena (average 22.36) than in DP1 (18.92). The professionals reduced the number or phenomena across sets, but they spotted an average of one phenomenon of high coincidence more in CT4 (3.65) than in DP3 (2.81). The teachers identified more singular phenomena in non-specialised texts. Across sets, they spotted fewer and fewer phenomena of INTERMEDIATE coincidence (very low, low and medium coincidence); their phenomena of high coincidence showed a similar pattern, except that CT4 had a slightly higher average (6.1) than DP3 (4.44). Lastly, the students used to intervene less in technical sets but, with regard to the phenomena of high coincidence, their behavior is close to that of the teachers.

	Singular				*Intermediate*				*High*			
	DP1	CT2	DP3	CT4	DP1	CT2	DP3	CT4	DP1	CT2	DP3	CT4
Add.	18.9	61.0	22.4	20.7	202.4	197.1	98.0	78.6	25.2	18.7	2.3	11.3
Prof.	71.9	34.1	25.6	21.0	520.1	164.1	91.0	89.7	44.8	9.2	2.8	3.6
Teach.	42.1	19.6	23.6	21.0	248.5	122.9	110.4	68.1	41.6	8.0	4.4	6.1
Stud.	58.7	27.2	42.4	22.1	402.1	176.1	193.1	148.6	45.1	12.1	4.7	8.0

Table 6. Phenomena saliency per group.

Table 7 displays quality judgments per set and group. If the total average had to be halfway between the second (bad) and the third (good) quality levels – therefore, at 2.5 –, then two of the four sets (DP1 and CT4) were (slightly) below average, and the other two (CT2 and DP3) were (also slightly) above average. The total average for the entire task was approximately 2.51.

	DP1	*CT2*	*DP3*	*CT4*
Total average	*2.40*	*2.61*	*2.60*	*2.51*
Addressees	2.71	2.39	2.83	2.47
Professionals	2.29	2.62	2.44	2.46
Teachers	2.32	2.79	2.84	2.50
Students	2.38	2.64	2.55	2.49

Table 7. Quality judgment per set and group.

The social science addressees were more lenient than the science addressees; the professionals were more demanding when evaluating non-specialised sets, whereas the other two groups involved in the translation sphere had worse general opinions on the translations of DP1 and, to a lesser extent, CT4.

4. Discussion

The number of actions in non-specialised sets almost doubled that in technical sets. But, when considering the results per set (Graph 1), the number of actions plummeted between the first and the following sets, probably because subjects learned in DP1 to evaluate series of translations. In any case, there were more actions also in DP3 than in CT2 and CT4; hence, subjects may have felt in general less self-confident to correct aspects within technical texts.

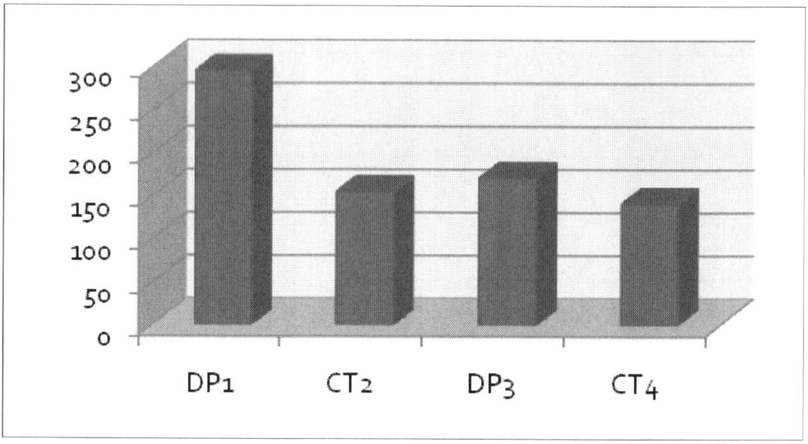

Graph 1. Actions per set.

The two professional groups reduced the number of actions across sets (Graph 2), and the students turned to be the subjects responsible for the upward trend observed in DP3. As a result, the aforementioned lack of confidence to evaluate technical texts may be due basically to the translation students. Besides, teachers show a slighter fall in the number of actions between DP1 and CT2, perhaps as a consequence of this group being more used to serial evaluation or, to put it simply, they did not need the first set – as much as the other groups – to adjust the level of demand and effort.

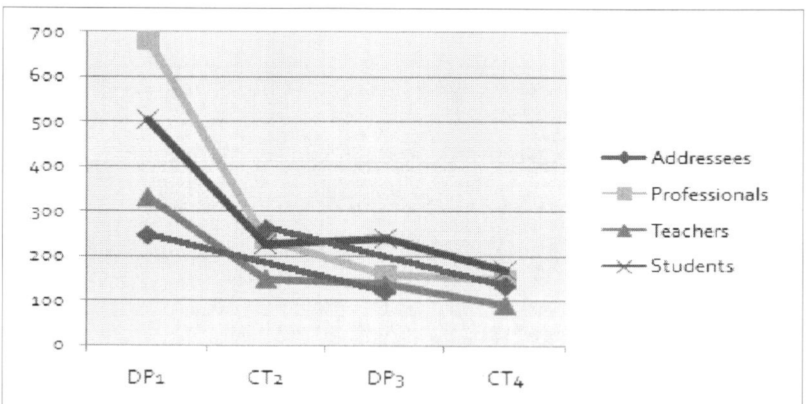

Graph 2. Actions per set and group.

Across sets, feedback-oriented changes were better balanced than product-oriented ones. Graph 3 also shows that the percentage of product-oriented changes was much higher in technical sets. This result may be taken to indicate that subjects did not feel qualified enough to change the texts because they mastered neither the subject matter nor the scientific language.

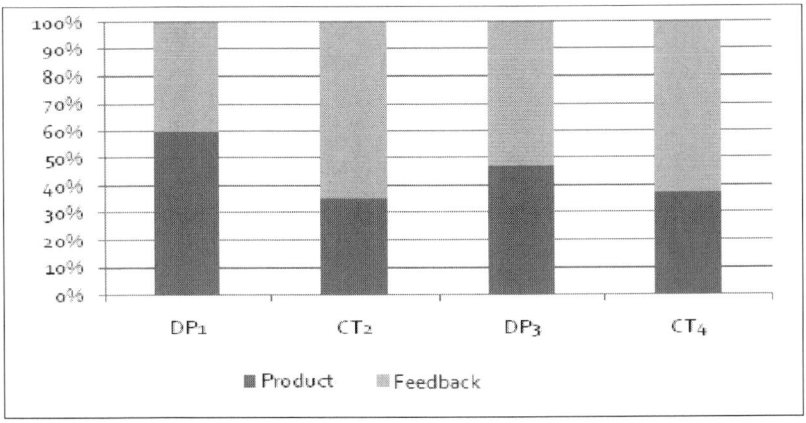

Graph 3. Percentage of changes in each set, according to their orientation.

Graph 4 shows the results per group. There was a general tendency to reduce the number of product-oriented changes across sets. In addition, technical sets show a higher percentage of feedback-oriented changes, mainly because the science addressees did not perform any product-oriented changes, but also because the students and, most especially, the professionals behaved in a clearly different way with regard to this type of changes, depending on whether the sets were specialised or non-specialised.

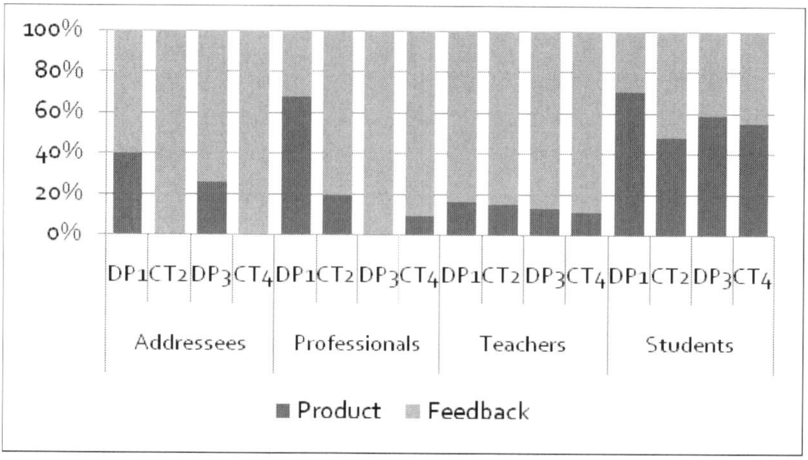

Graph 4. Change orientation per set and group.

The science addressees did not try to modify the text, but just to evaluate it. The professionals, who were used to leaving texts ready for use, stopped introducing product-oriented changes after DP1, which may be taken to indicate that they got tired of the task (they were not experts in serial evaluation). Nevertheless, they did not stop performing product-oriented changes on technical texts (unlike on DP3): they probably were extra-motivated by the difficulty of the subject matter, so that they tried to improve some fragments. The teachers introduced few product-oriented changes, possibly because their goal was to provide feedback for the student-translator, rather than to improve the text. Besides, as they were more used to serial evaluation, their behavior was more constant across sets. In CT2, the students performed more feedback-oriented than product-oriented changes: they did not modify these translations as much as the ones in non-specialised sets, probably because they did not feel so confident with the subject matter.

Graph 5 shows the percentages of revision and investigation comments but, to underline the differences, only between 50 and 100% (the rest were also revision comments).

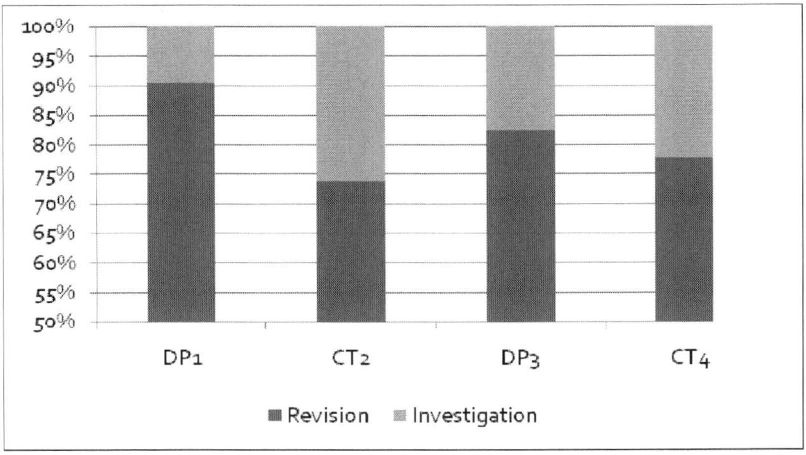

Graph 5. Comment contribution per set.

Revision comments were, clearly, the most common in every set. There was, however, a jump of investigation comments on technical texts. Therefore, subjects may have suspected there was an error but did not know the correct solution to it, so that they would have addressed the researcher – or, simply, introduced general assessments – to hide their doubts.

Graph 6 shows that the science addressees performed more revision and investigation comments than the social science addresses (who were not as many as the others). But, across sets, there was a similar evolution in the amount of comments, regardless of their type: the number of comments went down between technical sets, and up between non-specialised sets.

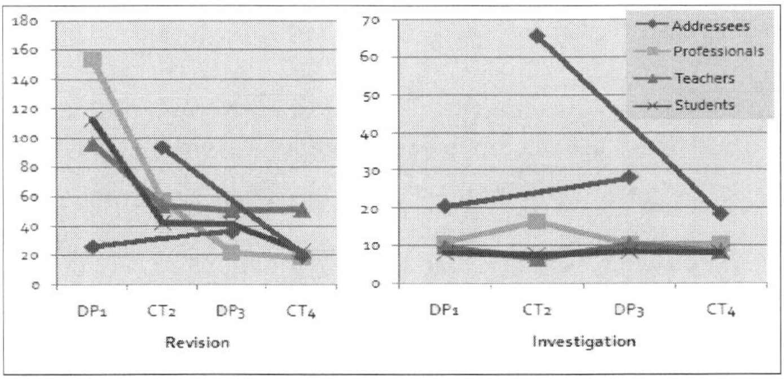

Graph 6. Comment contribution per set and group.

The three groups involved in the translation sphere show a steady decrease of revision comments, probably because these comments entailed a bigger effort. The investigation comments were better balanced across sets, but the professionals show a remarkable surge in CT2. The first technical set might have made them insecure and, as a consequence, they would have started addressing the researcher or assessing different aspects of the text, without correcting it. The aforementioned effect ascribed to the subject matter – there were more investigation comments on technical sets – might be attributed, as well as to professionals, to the science addressees, who were more detailed than their colleagues, especially with respect to the investigation comments in CT2.

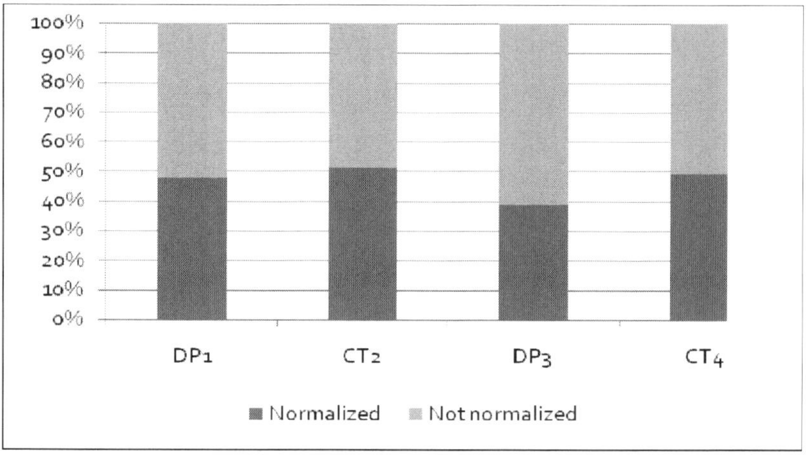

Graph 7. Nature of the phenomena per set.

Graph 7 shows the percentage of normalised and not normalised phenomena per set. The two types of phenomena were around 50%, except in DP3, where not normalised phenomena were clearly more common.

After DP1, where subjects marked many actions of any nature, normalised phenomena predominated in technical sets, surely because these phenomena depend on stylistic approaches that do not leave a big margin for nuances. Graph 8 displays the results per group.

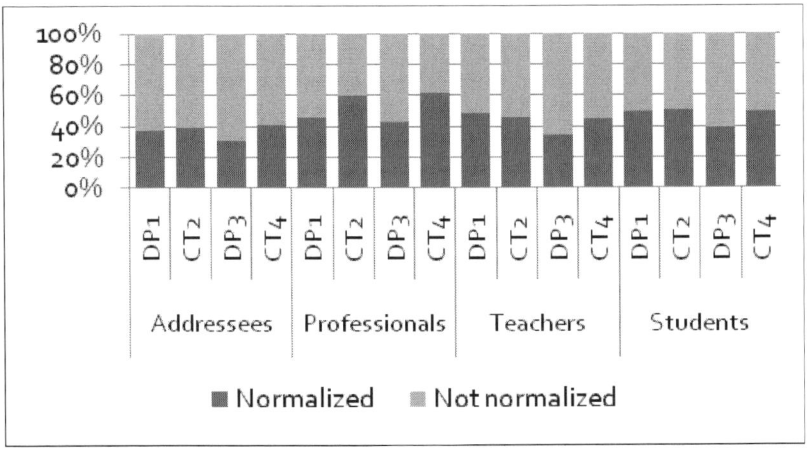

Graph 8. Nature of the phenomena per set and group.

The number of not normalised phenomena tended to increase in DP3. In non-specialised sets, subjects, regardless of their group, seemed to feel more entitled to correct not normalised aspects. This result may indicate either that subjects felt more comfortable with the subject matter, or that the language used in technical texts – more mechanical and less creative – made it somehow more difficult for arbitrary errors or phenomena to appear.

As for saliency, the phenomena of very low, low and medium coincidence were analysed together under the category of intermediate coincidence. Graph 9 shows the percentages of the three types of phenomena within each set.

The phenomena of intermediate coincidence had a similar percentage in the four sets. The other two categories show slight differences. On the one hand, the phenomena of high coincidence nearly disappeared from DP1 to DP3, whereas they were more constant in technical texts, possibly because these phenomena are about rules for writing numbers, conversion of weights and measures, terminology, etc. On the other hand, the number of singular phenomena increased in DP3, which supports the aforementioned result on not normalised phenomena in this set.

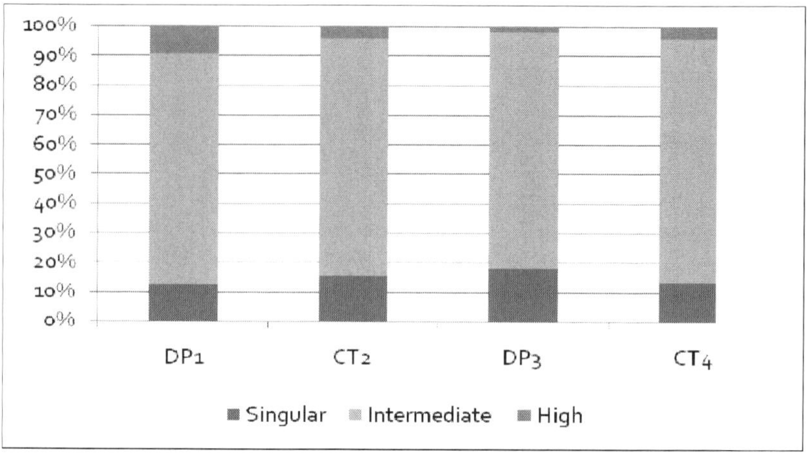

Graph 9. Saliency of the phenomena per set.

All groups show a growth of the singular phenomena (and a fall of the phenomena of high coincidence) between DP1 and DP3, but the gap was bigger among the social science addressees: they intervened less in this set than in DP1, thus coinciding also less than their colleagues with regard to the identification of phenomena. Within the teachers, Graph 10 shows that the percentage of singular phenomena was higher in CT4 than in CT2. Perhaps, when the teachers faced CT4, they felt more comfortable with the subject matter owing to the repetition (for originals CT2 and CT4 shared their second part). That is why they would have started performing a more personal evaluation, correcting not only the most evident errors.

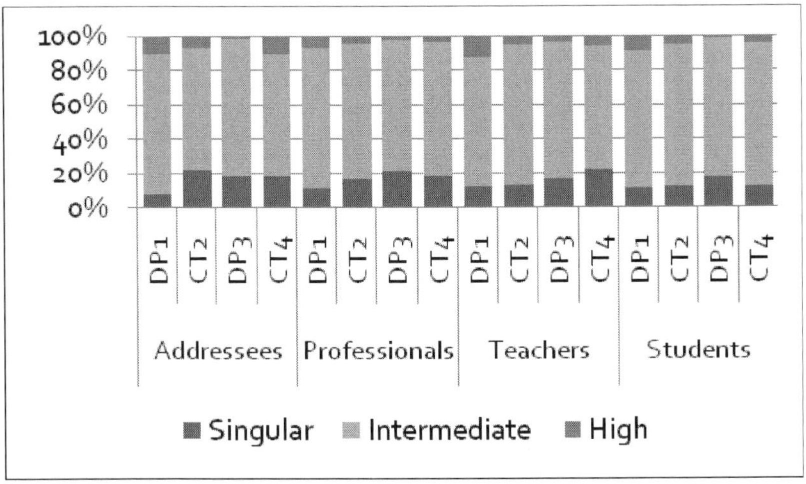

Graph 10. Saliency of the phenomena per set and group.

The last parameter was quality judgment. The set DP1 gathered the highest number of actions (*cf.* Table 1). As Graph 10 shows that the averages of the four sets were pretty similar, quality judgment appears not to correlate negatively with the amount of actions performed.

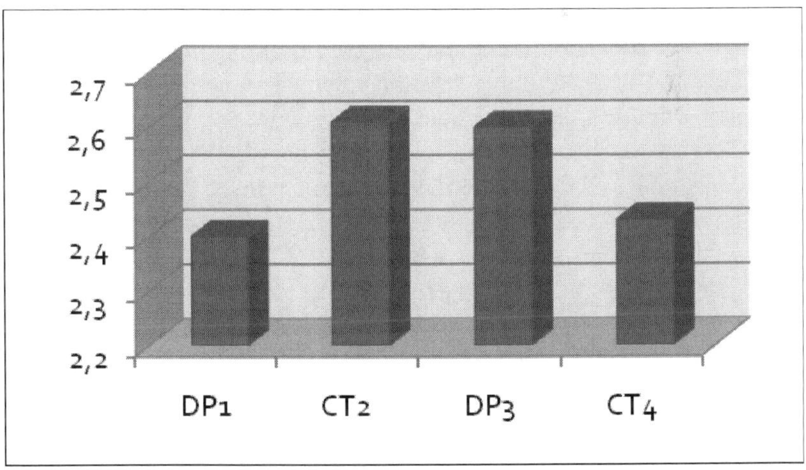

Graph 11. Quality judgment per set.

The subject matter does not seem either to have had an impact on quality judgment. Graph 12 displays the quality judgments assigned to each set by the four groups.

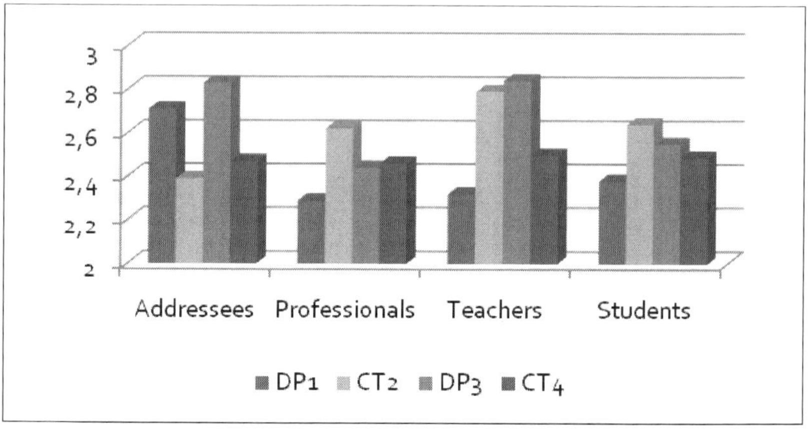

Graph 12. Quality judgment per set and group.

The science addressees were more demanding than their colleagues, the social science addressees, perhaps because they used to perform a more detailed evaluation and detected more errors. The professionals had better opinions on technical sets than on non-specialised sets. In contrast to DP1, where they performed many actions and to which they assigned a low quality judgment, the professionals deemed the first technical set much better, and that is why they gave better grades in CT2. This effect could be alternatively explained by their ignorance in relation to the subject matter. The teachers and the students assigned higher quality judgments to the central sets (CT2 and DP3). Further, the most demanding subjects, the professionals and the students, showed patterns which were especially similar. The students seemed to have interiorised the approach to quality that the profession would demand from them, whereas the teachers might have taken into account that translators who had carried out the texts were still students; hence, the translation teachers were generally more lenient.

5. Conclusions

The results make it possible to distinguish types of texts on the basis of their subject matter. There were two aspects which seem to have affected the differences observed in the evaluation of technical and non-specialised texts: (1) the evaluators' self-confidence about the subject matter and (2) the nature and disparate complexity of the technical and the non-specialised prose.

Subjects felt more comfortable and certain with non-specialised sets because they mastered the subject matter. That is probably why technical texts gathered fewer actions but, at the same time, more feedback-oriented changes and more investigation comments. The feedback-oriented changes and the investigation comments are alike in that they do not straightforwardly create a new text; therefore, if they were more frequent in technical texts, subjects might not have been convinced of their own capability to solve the errors that they spotted, and might have preferred to draw the attention away from themselves, for instance, by addressing the researcher. The biggest amount of actions in non-specialised sets was especially due to the translation students; the broader usage of feedback-oriented changes in technical sets, to the addressees, the professionals and the students and, finally, the higher percentage of investigation comments in technical sets (specifically in CT2), to the professionals. With regard to these parameters, the behavior of the teachers was clearly more regular, surely because they were the only group used to evaluating series of translations.

The structures and the language used in non-specialised texts were generally less restricted than that in technical texts, where there was less room for variation and creativity. The results of this study seem to suggest this tendency: on the one hand, not normalised phenomena were more often spotted in non-specialised sets and, on the other hand, singular phenomena were more frequent in DP3 than in technical sets. The (freer) language of non-specialised texts fostered the possible discordances between translator and evaluator as regards taste matters (not normalised), which, besides, were less likely to be shared by several evaluators; hence, there were fewer singular phenomena in technical sets. These results were found in the four groups: the potential addressees, the professional translators, the translation teachers and the translation students.

Lastly, the quality judgment, unlike in Conde (2007), did not seem to correlate directly with the subject matter. Nonetheless, subjects tended to raise the quality judgment between the first and the second of the non-specialised sets, whereas the average judgment declined between CT2 and CT4. The former result may be explained by the subjects learning to separate quality from quantity, for DP3 – which contained many actions – was given a higher quality judgment than DP1. As for the other result, subjects assigned higher judgments to the second technical set, once they had gained self-confidence in the subject matter. This happened in all groups, except among the addressees, who were the only experts in the subject matter and, for that reason, the self-confidence effect did not apply.

To sum up, the science addressees' main feature, apart from that of assigning a lower quality judgment to the second technical set, was that they introduced more comments than their colleagues (the social science addressees); as regards the professionals, they resorted to the feedback-oriented changes and the investigation comments whenever they did not master the subject matter; as regards the teachers, they showed a more regular evaluation across sets and as regards the students, they stopped modifying the text and, overall, started intervening less when they did not feel comfortable with the subject matter.

The effects analysed in the present study may be of interest because, aside from describing the originals through the evaluation of their translations, they call attention to possible tendencies and relations among the variables. Of course, there is still much more room for improvement, and future research could vary the evaluators' number or profile, as well as alter the order of the translations so as to confirm or reject the order effects and, consequently, draw generalisations useful for the community that works with texts of different nature.

References

Adab, Beverly 2002. The Translation of Advertising: A Framework for Evaluation. *Babel*. 47/2, 133-157.

Campbell, Stuart J. / Hale, Sandra 2003. Translation and Interpreting Assessment in the Context of Educational Measurement. In Anderman, Gunilla M. / Rogers, Margaret (eds) *Translation Today*. Clevedon: Multilingual Matters, 205-225.

Darwish, Ali 2001. Transmetrics: A Formative Approach to Translator Competence Assessment and Translation Quality Evaluation for the New Millennium. <http://www.translocutions.com/translation/transmetrics_2001_revision.pdf>

Klaudy, Kinga / Károly, Krisztina 2000. The Text-organizing Function of Lexical Repetition in Translation. In Olohan, Maeve (ed.) *Intercultural Faultlines. Research models in Translation Studies I. Textual and Cognitive aspects*. Manchester: St. Jerome, 143-160.

Muñoz, Ricardo / Conde, Tomás 2006. Effects of Serial Translation Evaluation. In Schmitt, Peter A. / Jüngst, Heike E. (eds) *Translationsqualität.* Frankfurt: Peter Lang, 428-444.

Muzii, Luigi 2006. Quality Assessment and Economic Sustainability of Translation. *Rivista internazionale di tecnica della traduzione*. 9, 15-38.

Orozco, Mariana 2001. Métodos científicos en traducción escrita: ¿Qué nos ofrece el método científico? *Sendebar*. 12, 95-115.

Reiß, Katharina 1971. *Möglichkeiten und Grenzen der Übersetzungskritik*. Munich: Max Hueber.

Riccardi, Alessandra 2002. Evaluation in Interpretation. Macrocriteria and Microcriteria. In Hung, Eva (ed.) *Teaching Translation and Interpreting 4. Building Bridges*. Amsterdam/Philadelphia: John Benjamins, 115-126.

Schiaffino, Riccardo / Zearo, Franco 2001. The Measurement of Quality in Translation. Presentation for the 42nd ATA Annual Congress, Los Angeles. <http://http://www.translationquality.com/downloads.htm>

Pilar Ordóñez López

The GENTT *Corpus of Specialised Genres:* A Valuable Tool for Professional Translators

1. Introduction

Textual genres, defined as "conventionalised forms of texts which reflect the functions and goals involved in particular social occasions as well as the purposes of the participants in them" (Hatim/Mason 1990: 69), constitute a key element in the translating process within specialised communication.

The concept of textual genre is the core element of an empirical investigation covering three fields of specialisation – legal, medical and technical – that is currently being carried out by the research group GENTT (Géneros textuales para la traducción/Textual Genres for Translation) in the Department of Translation and Communication at the Universitat Jaume I in Castellón de la Plana (Spain)[1].

Departing from the definition of genre provided by Hatim/Mason (1990), the GENTT project conceives genre as a dynamic category that changes with the socio-professional and cultural parameters within each field of specialisation (García-Izquierdo 2005). As such, genres represent an interface between text and context, both source and target (Montalt 2003), which enables translators to tackle the distance between their position as outsiders (García-Izquierdo/Montalt 2002) and the specialised professional frameworks the texts belong to. Through textual genres, translators become aware of and familiar with the socio-professional conventions, both linguistic and extra-linguistic, of different fields within specialised communication. Thus, the concept of genre

---

1 Research project funded by the Spanish Ministry of Education, 2006-09 (HUM2006-05581/FILO).

within the GENTT research group is three-dimensional, combining aspects from systemic functional linguistics, the theory of genre, and the sociology of professions; in other words, genre is considered to be an interdisciplinary category. The interdisciplinary nature of genres, as viewed by the GENTT group, makes it possible to combine the main components of specialised knowledge with those defining the translating process. In addition to the thorough theoretical investigation carried out since the year 2000, GENTT's investigation also has an important practical side, derived from the empirical-descriptive study of genres. This practical application is of special relevance within the professional framework of specialised languages.

This article aims to provide a revision of the characteristics of specialised communication, placing particular emphasis on legal translation. My purpose is to illustrate the usefulness of the GENTT *Corpus of Specialised Genres* by analysing the role of this tool at each stage of the translating process, as performed by a specialised legal translator.

2. GENTT within the framework of specialised communication

Specialised translation is, necessarily, an act of specialised communication. As mentioned above, GENTT's investigation covers three different fields of specialisation: legal, medical and technical. Each of these fields presents its own characteristics which go beyond purely linguistic aspects, such as terminology, fraseology, etc. Specialised communication is also characterised by extra-linguistic aspects, such as cognitive and communicative elements, as Cabré (1993: 139) states:

> Los lenguajes de especialidad son subconjuntos del lenguaje general caracterizados pragmáticamente por tres variables: la temática, los usuarios y las situaciones de comunicación.[2]

---

2 Our translation: "Specialised languages are subsets of general language, pragmatically characterised by three variables: subject, user and communicative situation".

Amongst the numerous studies examining specialised languages and how these are related to natural languages (e.g. Schifko 2001; Cabré 2002; Cabré/Estopa 2005), those which acknowledge the role of textual genre as a fundamental component of specialised languages and texts are especially relevant to GENTT, as pointed out by García-Izquierdo (2007). Indeed, the three variables included in the above definition of specialised languages (Cabré 1993) are embedded in GENTT's concept of textual genre. According to the theory underlying GENTT, genres play a crucial role in the configuration of specialised languages and therefore, textual genre is a key tool in the analysis of specialised communication. Similarly, the concept of a 'system of genres' (Bazerman 1994) allows us to understand the conceptual interrelations established both within a field of specialisation and between the different specialised fields.

Furthermore, regarding translation, the specialised translator is positioned as an outsider (Montalt 2003) in comparison to the specialists in the field and, to a certain extent, frequently regarded by them as an 'intruder'. In this sense, textual genre constitutes an important instrument of socialisation (Monzó 2003), enabling the translator to focus not only on linguistic and functional aspects, but also on the cultural conventions characterising each genre ('genre conventions') that specialised professionals in any area share. The fact that genres are categories shaped by cultural and professional conventions makes them useful tools in the process of socialisation undergone by the specialised translator:

> Este conocimiento de género o competencia de género multilingüe y multicultural no sólo define la experiencia del traductor ni su identidad profesional, sino que también es crucial en la práctica para que actúe como comunicador interlingüístico e intercultural. (GENTT 2005: 12)[3]

3 Our translation: "This knowledge or multilingual and multicultural competence not only shapes the translator's experience and professional profile, but it is also crucial, on the practical front, for the translator to act as an interlinguistic and intercultural communicator".

The GENTT research group is constructing a multilingual specialised corpus based on textual genre, which will be described in the following sections. The use of genres as a basis for the creation of a knowledge management system in the area of specialised communication offers important benefits for translators and writers of specialised texts; working with the concept of genre makes it posible to create "a mixed knowledge management system which combines translation-related knowledge with electronic management and information recovery systems"[4] (Borja 2005). The development of this multilingual specialised corpus is underpinned by GENTT's genre trees, which have been designed following an empirical investigation and with the assistance and collaboration of professionals in the three above-mentioned specialised fields. The genre trees constitute a structured, interrelated and dynamic conceptual instrument for specialised translators. Moreover, the development of the GENTT corpus is an 'open process' in continuous evolution, allowing it to incorporate socio-professional changes and thereby taking into account the dynamic nature of each field of specialisation.

3. Specialised translation: legal translation

In broad terms, 'specialised translation' can be defined as translation in which the translator is required to have certain specialised knowledge in a particular field in order to perform his/her task. Traditionally, research on specialised translation has focused almost exclusively on the study of terminology. However, in recent years the relevance of taking into account other aspects, such as the thematic field and genre, has been recognised. The following table contains Gamero's (1998) characterisation of specialised texts as well as the skills required to work as a specialised translator.

4 Our translation.

CARACTERÍSTICAS DE FUNCIONAMIENTO TEXTUAL	COMPETENCIAS REQUERIDAS DEL TRADUCTOR	
Importancia del campo temático	Conocimientos temáticos	CAPACIDAD PARA DOCUMENTARSE
Terminología específica	Conocimientos de terminología	
Géneros característicos	Conocimiento de los géneros característicos	

Table 1. Characteristics of specialised translation (Gamero, 1998: 100)[5].

Today, the increasing specialisation and segmentation of the market demands a higher degree of specialisation from translators. It would be utopian to think that this high degree of specialisation is reached by translators during their academic training. It is therefore fundamental that, within the area of specialised translation, translators have the necessary skills to easily access the information they need, in order to broaden or consolidate their thematic and terminological knowledge. In GENTT's approach, genre is a key element in the transfer of this knowledge, which is made easily accessible in the GENTT corpus. As a result, the documentation stage, comprising repetitive tasks such as the search for glossaries and parallel texts, becomes more dynamic and efficient, making a comprehensive sample of specialised genres and subgenres of each field of specialisations contained in the corpus available to the translator.

Naturally, the general characteristic features of specialised translation identified above are also applicable to legal translation. Nevertheless, as any other specialised field, legal translation has a set of defining features of its own.

5 Our translation: "Characteristics related to textual aspects: Importance of thematic field, specialised terminology, most common genres. Skills required to work as a specialised translator: thematic knowledge, terminological knowledge, knowledge of specific genres; skills to access the necessary resources and information".

> Podríamos proponer una primera definición de la «traducción jurídica» como la traslación de una lengua a otra de los textos que se utilizan en las relaciones entre el poder público y el ciudadano [...] y también, naturalmente, de los textos empleados para regular las relaciones entre particulares con trascendencia jurídica [...]. (Borja 2000)[6]

This is only a general definition and, as Borja (2000) acknowledges, refers only to the functionality of legal texts. In order to carry out an in-depth description of legal translation, Borja (2000) uses the classification model proposed by Hurtado Albir (1996), which consists of five variables: the function of the original text, the translation mode (which does not necessarily have to coincide with the original mode), the nature of the translation's function (distinguishing between professional and utilitarian translation), the direction of the translation, and the method of translation. Using these five variables, Borja (2000) carries out a thorough analysis of the main features of legal translation.

The function of the original text is analysed taking into account its thematic field, mode, tone and function. The thematic field of legal translation is law, even though, as Borja (2000) notes, overlaps can be found. For instance, Borja (2000) identifies a strong overlap between legal texts and commercial and administrative documents, "to the extreme that the traditional distinction between legal and comercial translation is often unclear"[7] (Borja 2000). Regarding their mode, legal texts can be oral, written, written to be read aloud, and oral to be recorded; their tone is, generally, extremely formal, with archaic elements; language functions are combined in legal texts, making them 'multifunctional texts' (Borja 2000), although the most common functions are instructive and exhortative. The translation mode is usually written (Borja 2000); however, sight translation and interpretation are also common. The nature of the function of the translation is normally one that can be defined as professional translation, according to Hurtado Albir's (1994) distinction, meaning that translation is an end in itself. Translation into

6 Our translation: "A first definition of 'legal translation' might be the transferral, from one language into another, of texts dealing with relations between public powers and citizens [...] and, naturally, also texts with legal implications, used to regulate relations amongst citizens".

7 Our translation.

the translators' mother tongue is recommended (Borja 2000), although this recommendation is not always followed in practice. Regarding the method of translation, Borja's (2000) identifies an eclectic situation: "translators choose one method or another, depending on the function of the translation".[8]

In addition to the previous characterisation of the field of legal translation, two main aspects must be taken into account as they play a defining role in the nature of legal translation. On the one hand, the conceptual and formal complexity of the thematic field is an important feature.

> La complejidad del lenguaje jurídico no proviene sólo de la gramática, sino que es consecuencia de los aspectos pragmáticos que lo contextualizan; no puede atribuirse únicamente a factores lingüísticos; sino a la combinación de una estructura conceptual compleja que impone una forma de expresión sofisticada. (Borja, 2000)[9]

On the other hand, given that translation is an intercultural activity, the lack of equivalence or correspondence between the two legal systems must be taken into consideration. This lack of equivalence presents a further difficulty for the translation and "can only be solved by an in-depth knowledge of the legal systems [...] that the translation brings into contact" (Borja 2000)[10].

4. The GENTT corpus of specialised genres

The compilation of the GENTT corpus was initiated in the year 2000, with the aim of providing specialised translators with a useful working tool which, at the same time, represents the practical application of

8 Our translation.

9 Our translation: "The complexity of legal language is not only due to grammar, but it is the result of the pragmatic aspects that contextualise it; this complexity cannot be attributed solely to linguistic factors, but to a combination of conceptually complex structures which impose a sophisticated way of expression".

10 Our translation.

the theoretical investigation undertaken by GENTT on the concept of genre. The GENTT corpus currently includes more than 900 textual occurrences, which is equivalent to more than two million words. In the creation of this corpus, an empirical-descriptive methodology was applied. The identification and compilation of each genre and subgenre is based on the analysis of "the function of the cultural context, the situational context and the existence of a differentiated linguistic materialisation"[11] (Borja 2005: 16). The GENTT corpus of specialised texts has the following features:

- The GENTT corpus is a specialised corpus, covering three fields of specialisation: legal, medical and technical.
- It is a multilingual corpus. The GENTT corpus contains texts written in five languages: Spanish, Catalan, English, German and French.
- It is a textual corpus, which implies that complete texts are included.
- The GENTT corpus is a dynamic compilation. As mentioned above, the design of the genre trees is an open process, in constant development in order to incorporate the socio-professional changes experienced by each field of specialisation.
- The GENTT corpus is a synchronic compilation.
- It is an online corpus. It is possible to access the corpus on the Internet. It is currently not yet possible to access the entire corpus, due to issues related to the copyright of the texts it contains. It should be borne in mind that some of the documents included in the corpus are of confidential nature; for instance, some of the legal texts are contracts, and some of the technical documents deal with patents. At the moment, it is nevertheless possible to access a provisional demo version on the GENTT webpage. In the near future, access to all parts of the corpus will be possible, with the restriction of it being used exclusively for academic purposes.
- It is a comparable corpus. The GENTT corpus is a compilation of original documents in five languages as well as a small number of translations. Comparative corpora make it possible to

11 Our translation.

compare genres in the different languages, which is of great use for translators. Ultimately, the GENTT group aims to transform it into a parallel corpus, so that translated versions of the original texts it contains are also included.

- It is an annotated corpus, as far as headers are concerned (see graph 1 below).

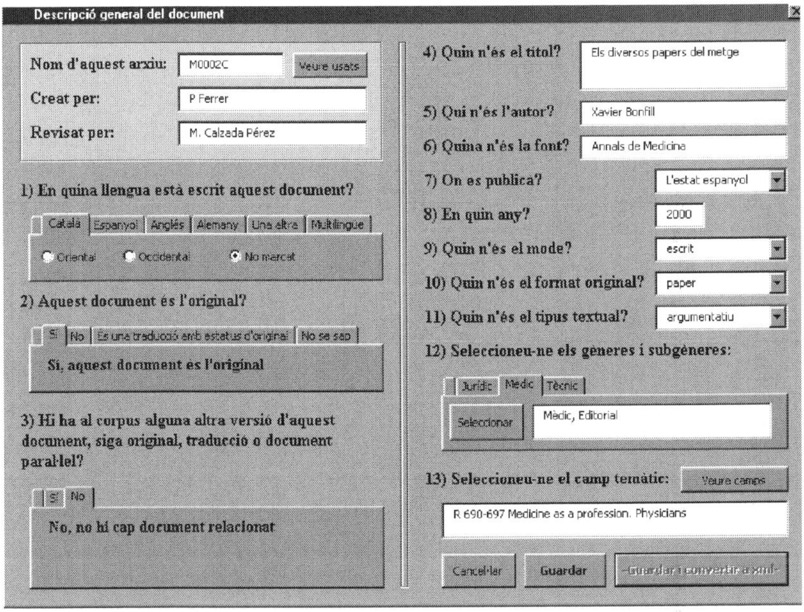

Graph 1. GENTT's header.

Each document has a header which provides the following information: the language the texts are written in; whether the document is original or translated; whether there are any other versions of the document included in the corpus, either translated versions or parallel texts; the author, year and place of publication; the mode and text type (argumentative, instructive or expositive); format; genre and subgenre (if applicable); field, according to the classification established by the Library of Congress; the name of the researcher who has added the document into the corpus, as well as the name of the person who has revised this process.

- Finally, the GENTT corpus is a 'user-friendly' tool. The GENTT corpus management tool (Jennings 2003), designed by one of the researchers of the group, makes it possible to use the corpus in a very straightforward and intuitive manner.

Using the GENTT corpus management tool, it is possible to obtain information about the texts included in the corpus through the GENTT search engine. The search engine allows for searches within the following criteria: title, genre, thematic field, language, original/translated documents, author, translator, source, place and year of publication, and text type. New applications have recently been incorporated in order to expand the usefulness of the corpus for research purposes. The study of collocations, frequencies and phraseology can now be undertaken within the GENTT corpus, as shown in the demo version on the GENTT website. To date, no body mark-up or lemmatization has been carried out, but it is currently being considered how elements of annotation can best be incorporated.

The only field of specialisation for which two separate genre trees have been developed is that of legal texts. This is due to the above-mentioned lack of correspondence between the continental legal system and the Anglo-American one.

5. The GENTT corpus in the translating process

As mentioned above, the GENTT project is based on the assumption that textual genre is a relevant instrument for the analysis of specialised communication and, thus, it is a valuable tool in the area of specialised translation. Specialised translation is an act of specialised communication between two different languages and two different socio-professional communities. The GENTT corpus constitutes an extremely useful resource for specialised translators. Furthermore, this corpus can be used for translator training and members of the GENTT group have, in fact, been using it in their translation courses for a long time now, obtaining very satisfactory results. In this article, the focus lies on the professional use of the GENTT corpus by specialised translators working in the legal field; the aim is to analyse the role played by the corpus at each stage of the translating process.

5.1. The translating process

The translating process is defined as "the mental process which allows the transmission of a text written in a particular language, using the means of a different one" (Hurtado Albir 2001: 640)[12]. First of all, it should be noted that the study of the translating process is not a simple task, as it involves analysing the mental processes perfomed by a translator when he/she is translating. Mostly due to this intrinsic difficulty, previous approaches by various scholars, some of which will be reviewed below, generally lack solid empirical evidence.

Hurtado Albir (2001: 311-374) presents and reviews several models suggested for the description of the translating process, as proposed within various approaches: the 'interpretative model', proposed by the *ESIT*, the linguistic and the psycholinguistic models by Bell, and the socio- and psycholinguistic models by Kiraly, amongst others. A number of similarities can be observed between these models, such as the distinction of two essential stages: (a) understanding or analysis (Bell 1991) and (b) rewriting or synthesis (Bell 1991). Some authors also differentiate a third, intermediate stage, that of 'deverbalisation' (Seleskovitch/Lederer 1984, 1989).

During the first stage of the translating process, translators are engaged in understanding the text, trying to confer sense upon it by employing their previous knowledge. The deverbalisation phase implies the construction of a referential image, deverbalised, which serves as a bridge between the source and the target text. The deverbalised concepts are then linguistically re-expressed in the target language. Finally, in the rewriting or synthesis stage, the translator rewrites the message of the source text in the target language, using his/her linguistic and thematic knowledge.

12 Our translation.

According to Hurtado Albir, in these models the translating process is characterised by the following features:

> 1) la existencia de procesos básicos; 2) el papel de la memoria y los conocimientos almacenados por ella; 3) el carácter interactivo y no lineal del proceso; 4) la existencia de procesos controlados y no controlados; 5) su consideración como un proceso de reconocimiento y de resolución de problemas, toma de decisiones y de utilización de estrategias; 6) la existencia de características específicas en cada modalidad y tipo de traducción, y según el método elegido. (Hurtado Albir, 2001: 367)[13]

Hurtado Albir (1996: 64-65) also provides a model of analysis of the translating process, combining aspects taken from other proposals, such as Delisle's (1982) interpretative model, Bell's (1991) informative model, and Hatim/Mason's (1990) model based on discourse analysis. Hurtado Albir's comprehensive approach is reflected in the following model (see graph 2), incorporating elements typical of discourse analysis which are extremely useful to break down into separate stages the process performed, consciously and/or unconsciously, by the translator. Within each of the three stages of the translating process identified above, Hurtado Albir further distinguishes four separate levels: the morphosyntactic, the semiotic, the pragmatic and the semantic ones; it is at these levels that the GENTT corpus is particularly useful in facilitating the translator's task.

13 Our translation: "1) the existence of basic processes; 2) the role played by memory and the knowledge it stores; 3) the interactive and non-linear nature of the process; 4) the existence of controlled and uncontrolled processes; 5) the consideration of the translating process as a process of recognition, problem-solving, decision-making and application of strategies; 6) the existence of specific characteristics for each translation mode and type and in accordance with the chosen method of translation".

Primera etapa	
Análisis morfosintáctico (Lingüístico)	Análisis semántico (Contenido)
Análisis pragmático (Contexto)	Análisis semiótico (Intertextualidad)
COMPRENSIÓN DEL TEXTO ORIGINAL	
Segunda etapa	
DESVERBALIZACIÓN	
Tercera etapa	
Sintetizador morfosintáctico (Lingüístico)	Sintetizador semántico (Contenido)
Sintetizador pragmático (Contexto)	Sintetizador semiótico (Intertextualidad)
REFORMULACIÓN DEL TEXTO DE LLEGADA	

Graph 2. Hurtado Albir's model of the translating process (1996: 64-65)[14]

5.2. The GENTT corpus in the translating process

After having examined the different stages of the translating process from a scholarly perspective, I will now adopt a position closer to the translator's approach, in order to illustrate the usefulness of the GENTT corpus in specialised translation. In this review of the role played by the corpus in the translating process performed by a specialised legal translator, the direction of the translation will be English to Spanish, which, in the case of native Spanish translators, will mean translating into their mother tongue. Three different stages can be distinguished: before the complete reading of the text, while reading the text, and after having read the text. These three stages will be analysed according to Hurtado Albir's description of the translation process.

14 Our translation: "First stage: morphosyntantic analysis (linguistic), semantic analysis (content), pragmatic analysis (context), semiotic analysis (intertextuality). Understanding the source text. Second stage: deverbalisation. Third stage: morphosyntactic synthesis (linguistic), semantic synthesis (content), pragmatic synthesis (context), semiotic synthesis (intertextuality). Rewriting the target text".

a) Before the complete reading of the source text

In legal translation, when a translator is presented with the source text for the first time, he/she tries to answer the question: 'What is it?', attempting to identify the document through the activation of his/her knowledge in the field. The GENTT corpus plays an important role as a body of reference at this stage, as the concept of textual genre allows a quick identification of specialised texts:

> [La noción de género] constituye el punto de encuentro de aspectos relacionados con el registro, la pragmática o la semiótica de textos, de vital importancia para un buen análisis textual, estadio imprescindible en el proceso traductor. (García Izquierdo, 2002)[15]

By introducing the title or genre of the document in GENTT's search engine (<http://sera0041.ahs.uji.es/recercar/search.php>), it is possible to obtain pragmatic and contextual information about its genre or subgenere. It is only possible to use any of the search criteria listed above (see section 4). Access to genre trees is, at the moment, restricted (<http://sera0041.ahs.uji.es/Informe.php?corpus=Juridic>). The Anglo-American genre tree provides the translator with a general overview of the conceptual organisation of the field. Furthermore, thanks to this genre tree, the translator can immediately categorise the source text and visualise the position occupied by it within the mapping of the legal field. Also, the genre tree makes it possible for the translator to identify the genre and/or subgenre of the text and recognise the different interrelations between the genre or subgenre the text belongs to on the one hand, and other possibly related genres on the other. These identification processes make it possible to perform the semiotic analysis (Hurtado Albir 1996).

Once the text has been identified, the translator can use the GENTT search engine in order to find similar texts. In the header annotation associated with each document, information about contextual aspects

15 Our translation: "[The concept of genre] constitutes a meeting point of aspects related to register, text pragmatics and text semiotic, which are fundamental components of a comprehensive textual analysis, an indispensable stage in the translating process".

can be found, which helps the translator carry out the pragmatic analysis of the text (Hurtado Albir 1996). The identification of the genre implies an automatic activation of the translator's conceptual, semiotic, linguistic and pragmatic knowledge. Moreover, access to documents belonging to the same genre as the original text enables the translator to identify the most characteristic, representative features of that genre, the genre conventions. The identification of genre conventions provides him/her with information about aspects such as the typical phraseology of each genre, which in turn facilitates semantic and morphosyntactic analysis (Hurtado Albir 1996).

The use of the GENTT corpus during the first stage of the translating process allows the translator to categorise the source text conceptually; this will help him/her to formulate, either consciously or unconsciously, a series of expectations, and to anticipate the problems the translation will face him/her with, thereby facilitating his/her understanding of the text.

b) During the reading

Once all the previous categorisations have been made and the translator has already built expectations about what he/she will find in the text, he/she starts reading it. While reading the text, he/she will be able to recognise the genre conventions identified previously in other documents included in the GENTT corpus (pragmatic analysis, Hurtado Albir 1996). Reading the text, the translator will identify the main ideas (semantic analysis, Hurtado Albir 1996) and recognise the textual structure (morphosyntactic analysis, Hurtado Albir 1996). These analyses contribute to a thorough understanding of the source text and allow the translator to detect examples of lack of equivalence, to anticipate solutions and to conceive a strategy to solve them. During this stage, the translator also makes decisions regarding the method of translation he/she is going to use on the basis of the analysed aspects, as well as on the function of the translation.

c) After reading the text

According to the model of translating process described above (see 5.1.), a stage of deverbalisation followed the understanding of the text.

During deverbalisation, the GENTT corpus plays an important role to help translators to build a conceptual bridge between the source and the target text, or between the source and the target genre. At the moment, the corpus does not include any referential information about the conceptual part of genres, i.e. no description or characterisation of genres. However, it is planned for conceptual information as well as bibliography resources for each genre to be included.

After the stage of deverbalisation, the synthesis or re-writing stage is performed. During this stage, the translator activates his/her linguistic and thematic knowledge of the corresponding legal system. GENTT genre trees provide the translator with a comprehensive mapping of the target legal system. The translator can retrieve the semiotic aspects (intertextuality) of the legal system in the target language, which allows him to categorise the text in the specialised target culture. This categorisation allows the translator to determine the degree of equivalence between the two genres involved. As total equivalence or a complete lack of equivalence is naturally not the only possibility, the genre tree also reflects partial equivalence and any nuances existing between the two legal systems. The translator can also access parallel documents, which helps to perform semantic and morphosyntactic synthesis (Hurtado Albir 1996). Through the GENTT header, the translator has access to contextual information regarding the target genre, provided that there are similar documents included in the corpus; this will provide the translator with the necessary information to carry out the pragmatic synthesis (Hurtado Albir 1996), as well as allowing him/her to identify the corresponding genre conventions.

The GENTT corpus brings a representative sample of legal genres and subgenres within the translator's grasp, saving him/her from having to undertake a laborious and repetitive documentary search, speeding up his/her task significantly. Moreover, thanks to the GENTT corpus management tool and the new search applications of concordance and frequency, in case of lack of correspondence between the legal systems involved, translators can use the corpus to extract information about formal aspects and macro-structure through the access to related genres and subgenres.

6. A practical case: the translation of a judgement-sentencia

In order to illustrate the process described above, the role of the GENTT corpus in a practical case of translation, the translation of a judgement, will be reviewed. For practical reasons, the example has been selected from those included in the provisional demo version that appears on the webpage of the GENTT research group. It should be borne in mind that in this demo version, only a small sample of texts has been included; working with the whole corpus will imply access to a more comprehensive sample. The selected text is 'Judgements - Macfarlane and Another v. Tayside Health Board'. The stages distinguished above will be followed in this revision.

a) Before the complete reading of the text

First, the translator can use the GENTT search engine (see Section 4). By entering the document's genre, it is possible to access all the documents in the corpus that belong to the same genre. If the translator is unsure about the exact genre of the text, the GENTT search engine offers other search criteria such as the title, which may be just as straightforward as the genre.

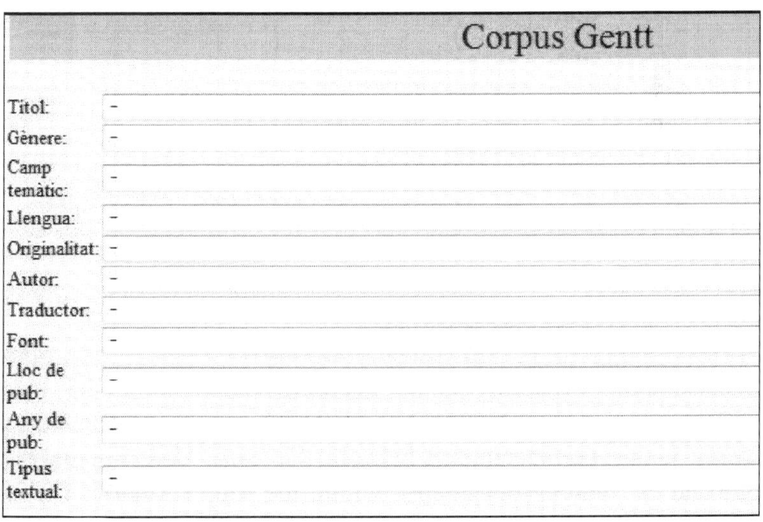

Graph 3. GENTT's search engine.

tree contributes to the conceptual categorisation of the genre: in this case, a 'sentencia' is categorised as a 'judicial document' within the continental legal system.

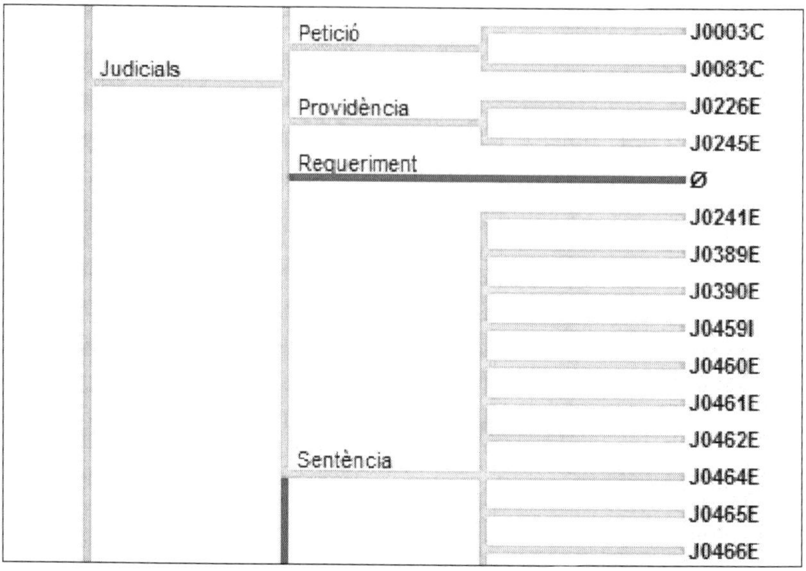

Graph 6. GENTT continental legal genres tree.

Either with the assistance of the genre tree or by using the GENTT search engine, the translator can establish whether there are any documents belonging to the same genre in the target language. The list of documents classified under 'sentencia' provides the translator with a valuable sample of documents from which genre conventions in the target language can be deducted. In this case, three documents are shown: 'Sentencia nº 418-97', 'Sentencia Tribunal Supremo (Sala de lo Contencioso-Administrativo, Sección 4ª)' and 'Sentencia del Tribunal de Primera Instancia. Laura Moretti contra Comisión de las Comunidades Europeas'. Being aware of genre conventions allows the translator to analyse the similarities or differences that exist between the source and the target language, both linguistically and extra-linguistically. Furthermore, in the list of documents, as mentioned before, contextual and pragmatic aspects of each document are provided.

7. Conclusions

The concept of textual genre provided by the research group GENTT incorporates the main features characterising specialised translation. Furthermore, the application of this concept to corpus linguistics has produced a valuable tool for specialised translators. As shown in this article, the use of the GENTT corpus in the translating process carried out by legal translators makes translation more dynamic, especially on the basis of access to specialised information. By giving the translator access to a representative sample of specialised texts, conceptually categorised in terms of genre, and by making information about the pragmatic and contextual aspects involved in them available to him/her, the GENTT corpus makes a genuine contribution to the field of specialised translation.

References

Bazerman, Charles 1994. Systems of Genre and the Enactment of Social Intentions. In Freedman, Aviva / Medway, Peter (eds) *Rethinking Genre*. London: Taylor and Francis, 79-101.

Bell, Roger 1991. *Translation and Translating: Theory and Practice*. London: Longman.

Borja Albi, Anabel 2000. *La traducción jurídica: didáctica y aspectos textuales*. Centro virtual Cervantes. Aproximaciones a la traducción. <http://www.cvc.cervantes.es/obref/aproximaciones/borja.htm>

Borja Albi, Anabel 2005. Organización del conocimiento para la traducción jurídica a través de sistemas expertos basados en el concepto de género textual. In García Izquierdo, Isabel (ed.) *El género textual y la traducción. Reflexiones teóricas y aplicaciones pedagógicas*. Bern: Peter Lang, 37-67.

Cabré, María Teresa 1993. *La terminología (teoría, metodología, aplicaciones)*. Barcelona: Antárdida.

Cabré, María Teresa 2002. Análisis textual y terminología, factores de activación de la competencia cognitiva en la traducción. In Alcina Caudet, Amparo / Gamero Pérez, Silvia (eds) *La traducción científico-técnica y la terminología en la sociedad de la información*. Castelló: Universitat Jaume I, 87-105.

Cabré, María Teresa / Estopa, Rosa 2005. Unidades de conocimiento especializado. Caracterización y tipología. In Cabré, María Teresa / Bach, Carme (eds) *Coneixement, llenguatge i discurs especialitzat*. Barcelona: IULA-Documenta Universitaria, 69-93.

Delisle, Jean 1982. *L'Analyse du discours comme méthode de traduction*. Otawa: University of Ottawa Press.

Gamero, Silvia 1998. *La traducción de textos técnicos (alemán-español). Clasificación textual: tipos y géneros*. Barcelona: Universitat Autònoma de Barcelona.

García Izquierdo, Isabel 2002. El género: plataforma de confluencia de nociones fundamentales en didáctica de la traducción. In *Discursos. Série Estudos de Tradução*, 2. Universidade Aberta: Lisbon.

García Izquierdo, Isabel 2005. El género y la lengua propia: el español de especialidad. In García Izquierdo, Isabel (ed.) *El género textual y la traducción. Reflexiones teóricas y aplicaciones pedagógicas*. Bern: Peter Lang, 117-134.

García Izquierdo, Isabel 2007. Los géneros y las lenguas de especialidad. In Alcaraz, Enrique (ed.) *Las lenguas profesionales y académicas*. Barcelona-Alicante: Ariel-IULMA, 119-125.

García Izquierdo, Isabel / Montalt, Vicent 2002. Translating into Textual Genres. *Linguistica Antverpiensia*. 1, 135-145.

GENNT 2005. El concepto de género: entre el texto y el contexto. In García Izquierdo, Isabel (ed.) *El género textual y la traducción. Reflexiones teóricas y aplicaciones pedagógicas*. Bern: Peter Lang, 7-15.

Hatim, Basil / Mason, Ian 1990. *Discourse and the Translator*. London: Longman.

Hurtado Albir, Amparo 1994. Modalidades y tipos de traducción. *Vasos Comunicantes*. 4, 19-27.

Hurtado Albir, Amparo (ed.) 1996. *La enseñanza de la traducción*, Castellón: Universitat Jaume I.

Hurtado Albir, Amparo 2001. *Traducción y traductología. Iniciación a la traductología*. Madrid: Cátedra.

Montalt, Vicent 2003. El gènere textual com a interfície pedagògica en la docència de la traducció cientificotècnica. In Cánovas, Marcos *et al.* (eds) *Actes de les VII Jornades de Traducció a Vic*. Vic: Universitat de Vic.

Monzó, Esther 2003. Las socializaciones del traductor especializado: El papel de los géneros. *Revista de la Facultad de Lenguas Modernas*. 6, 15-29.

Schifko, Peter 2001. ¿Existen lenguas de especialidad? In Bargalló, María *et al.* (eds) *Las lenguas de especialidad y su didáctica*. Tarragona: Universitat Rovira i Virgili, 21-31.

Seleskovitch, Danica / Lederer, Marianne 1984. *Interpréter pour traduire*. Paris: Didier Érudition.

Seleskovitch, Danica / Lederer, Marianne 1989. *Pédagogie raisonnée de l'interprétation*. Paris: Didier Érudition.

Francisca Suau-Jiménez

Metadiscursive Elements in the Translation of Scientific Texts

1. Introduction and research purpose: translation method and metadiscourse

Translation of scientific texts, as happens with other texts of specialisation, can be approached from different perspectives: discourse, register, genre, terminology, etc., as several authors have suggested (Bédard 1986; Cabré 1993; Maillot 1997; Gamero 2001; Montalt 2005). One successful approach is the pragmatic perspective that applies genre and register to translation (Suau-Jiménez 2001). This allows us to identify all communicative functions and translate them into the target text. In the same vein, register analysis permits us to identify the field and thus the whole lexical and semantic sphere of the text; the tenor tells us who the reader is and so how to address him/her in a meaningful and persuasive way; and, finally, the mode leads us to recognise and recreate syntax and lexical levels of specialisation. This translating method has yielded very good results: after a training period following a series of steps that lead translators to analyse genre and register, they automatise the process and identify generic structure and register characteristics in a short time. In this way translation is done in a faster and more secure way, since genre and register guarantee that structural, grammatical and lexical equivalences can be found easily

Until very recently, tenor only told us who the text sender and receiver were and thus, translators tried to adapt register to the addressee making language more or less formal but without following any given pattern or linguistic model. Metadiscourse, a recent applied linguistics concept coming from the development of the tenor of register and from the influence of semantics and the way to represent the phenomenon of the meaning of *hedging* (Lakoff 1972), has become an extremely useful new tool in specialised translation. It is described as the linguistic material that oral and written texts possess, not adding any propositional contents but meant to help the reader in the interpretation and evaluation of the message (Crismore *et al.* 1993). Metadiscursive elements such as hedges principally have been the object of analysis in scientific English (Salager-Meyer 1994) and thus the notion of metadiscourse has become a new concept itself, highlighting a pattern with a variety of linguistic resources within academic and scientific English (Hyland 1998, 2000; Hyland/Tse 2004). However, there are few contrastive works comparing metadiscursive elements in English and Spanish designed to be applied to specialised translation (Suau-Jiménez 2005a, 2005b, 2005c; Suau-Jiménez/Dolón 2007, forthcoming).

Both hedges in research articles and phatic elements in popular science are two examples of metadiscourse. They also constitute what Halliday (1978) calls *tenor* within the category of register, being a token of Halliday's *interpersonal function*. Departing from this theoretical principle I have analysed both types of resource – hedges and phatic elements – in a pilot corpus of research and popular science articles in English and Spanish. The purpose has been to account for those metadiscursive elements which are prototypical of English scientific discourse and see whether there is a correspondence with the Spanish scientific metadiscourse. Also whether the Spanish language and culture impose different ones, as happens in other specialised fields like tourism (Suau-Jiménez/Dolón 2007). The implications for translation are important since metadiscourse assists in the accomplishment of some important prescriptive functions in scientific genres: politeness and/or reader persuasion, both in research and popular science articles. If these texts are translated without considering metadiscourse of the target language and genre, one might unwittingly flout texts' communicative goal since persuasion would not be expressed as native speakers of the target language do and therefore, the result would be a deficient translation.

2. From register to metadiscourse in translation

The notion of metadiscourse comes from two fields of linguistic analysis, always from the discourse analysis perspective. On the one hand, it is the result of the evolution of the register concept (Halliday 1978) and, on the other, it derives from discourse analysis and semantics. It is known that register describes changes affecting language when this is used in a specific context, following field, tenor and mode. This tripartite subdivision corresponds to the ideational, interpersonal and textual functions that Halliday relates to ideas and concepts, the relationship between author and reader – phatic and politeness aspects, etc. – and the materialisation in the text itself – grammatical and lexical features. Since the time when Halliday coined the concept of register, it has been applied to the analysis and description of many oral and written examples in specific contexts. At the same time, different authors from the field of discourse analysis began to analyse elements that modulate and hedge language. *Hedges* were defined semantically by George Lakoff (1972), in order to complete Rosch's work (1973) in her prototype theory, as the elements that modulate or *hedge* meaning, thus assuming the vagueness and non assertiveness of language. Examples of hedges are the words and expressions *may control, could happen, would suggest, probably, possibly,* etc.

From the beginning, the study of hedges was linked to academic and scientific discourse, due to its need to use language in a tentative and non-assertive way. Soon, researchers of discourse modulation in English became interested in other textual elements apart from hedges: boosters (*in fact, definitely*), attitude markers (*I agree, surprisingly*) and engagement markers (*consider, note that*) (Hyland/Tse 2004): metadiscourse was born. From the pioneering work of Vande Kopple (1985) and Crismore (1989), a large number of studies followed (Crismore/Farnworth 1990; Crismore/Markkanen/Steffenson 1993; Mauranen 1993; ; Dahl 1994; Hyland 1998, 1999, 2000; and in Spain, Valero Garcés 1996 and Fuertes-Olivera *et al.* 2001), all applied to languages of specialisation. The work of Christiane Nord is also very substantial because of the same interest in the relation between author and reader applied to translation, although the name given to this phenomenon is not metadiscourse but metacommunication (Nord 2000, 2007).

Hyland/Tse (2004) propose a pattern for metadiscourse in academic and scientific English texts which is divided into two categories: textual and interpersonal. The latter clearly refers to the relation between author and reader, fundamental in performative and persuasive texts, that is to say, texts where the reader has to make a decision, to act or accept a new piece of information in the case of the scientific community. Following international guidelines for English written scientific articles in scientific journals worldwide, papers are only published if they accomplish certain principles like linguistic politeness, non-imposition of criteria and non-personalisation of the scientist's self (Alcaraz 2000; Suau-Jiménez 2005a). All these principles derive from the Anglosaxon cultural and academic pattern, now accepted worldwide in the field of science publishing in English. Therefore, the use of metadiscourse and its adequate translation, bearing in mind the differences between text of origin and target text is crucial for English-Spanish and Spanish-English translation, since metadiscourse varies from one language to the other and from one genre to the other (Suau-Jiménez 2006). However, studies on metadiscourse have not abounded in studying the differences among languages, genres and fields of specialisation, three parameters that should be combined for the sake of accurateness in the application of metadiscourse to specialised translation. Metadiscourse is not only important in the so-called 'serious science', i.e. research articles, but also in popular science, since it is also through entertainment and informal registers that readers are persuaded to accept new findings and theories. However, this kind of metadiscourse is radically different from that of research articles, in the same way in which its register is different to that of research articles (Suau-Jiménez 2005a). This is probably proof that metadiscourse is part of register and that it therefore shares its characteristics of formal or informal language elements, as will be further shown in this article.

3. Metadiscourse variation following language and genre

As I have mentioned before, bearing in mind the different metadiscourse patterns is essential when translating performative and/or persuasive texts, not only in research articles but also in popular science. Translation is a process that cannot only depend on the finding of equivalences in communicative functions and terminology. It is also necessary to take into account the interpersonal function, i.e. metadiscourse or the way in which the author addresses the reader. This is what varies from one language to another, from one genre to another and, possibly, from one area of specialisation to another. What Hyland/Tse (2004) describe is the English metadiscourse of scientific and academic texts, i.e. the formal register of science, which is very useful when teaching English academic writing. However, if one needs to translate those texts, either from English into another language or the other way round, it will be necessary to know in detail both types of metadiscourse: that of the text of origin and that of the target text, in order to adjust our translation. Otherwise we risk making deficient translations and obtaining a final product that 'creaks' to our ears because the metadiscourse elements are those of the text of origin instead of those of the target text, genre and field of specialisation.

With respect to metadiscursive differences among genres, there is an important one, for instance, between research articles and popular science. Their elements are totally different since both, the register and communicative purpose of these two kinds of texts, are also different. On the one hand, research articles intend to inform and persuade expert readers, specialists and scholars without imposing on them new theories or findings through language, as the scientific community requires. On the other hand, popular science articles attempt to inform and teach the general public through entertainment. To this end, language is used in a more informal way, its register being full of licenses such as metaphors, colloquial terms and phrases, plays on words, etc., with more categorical propositions than research articles, since there is no scientific community which has to accept the new findings. Readers must be convinced with strong arguments, not with tentative ones. These characteristics demonstrate that metadiscourse is a new concept in the analysis of language that is strongly linked and depends on register and genre at the same time, being the result of the communicative goal and of its necessary representation, register.

4. Metadiscursive differences in research articles and popular science texts in English and Spanish: a pilot study

I have carried out a pilot study with a corpus of research articles and popular science ones in English and Spanish (20 texts) in order to draw some general conclusions about metadiscourse and its differences in terms of language and genre. My purpose has been to find implications for translation of these two scientific genres in English and Spanish. The frameworks I have followed have been Hyland's (1998) for research articles and Suau-Jiménez's (2005a) for popular science ones:

4.1. Epistemic resources, hedges and modal verbs (Hyland 1998)

These include the following features:

- Modal verbs: *may, suggest, could.*
- Epistemic grammatical categories: *predict, suggest, possibility, relativeness, likely, presumably.*
    - Expressions that limit conditions: *"We have not been able to determine..."*
    - Expressions that admit lack of knowledge: *"Nothing is known about..."*

4.2 Phatic resources in popular science (Suau-Jiménez 2005a)

These have been classified as follows:

- morphosyntactic: *personal pronouns, imperatives.*
- lexical: *colloquial terms.*
- semantic: *meaning subjectivity.*
- stylistic: *calls to the reader, plays on words.*

Consider this example:

(1) *If you were asked* which is the little animal living in your garden that is both male and female, *what would you answer*? Snails, *isn't it*? *Great! Don't doubt it*.

5. Results

5.1. Characteristics of metadiscourse in research articles in Spanish

Impersonal forms / Discourse hedges

Examples:

(2) Actualmente *se asume* que el cáncer es el resultado de alteraciones en una célula que modifica su potencial de crecimiento, la respuesta a los mecanismos de control y función. Estos cambios *pueden implicar* mutaciones en uno o más genes, anormalidades cromosómicas graves o anomalías en la transcripción genética que conducen a desdiferenciación y alteración de las propiedades de crecimiento de la célula. (<www.forobioquimico.com>)

(3) *Se seleccionó* un grupo de estos molares con diferentes grados de calcificación radicular que iba desde: con trazo de formación de las raíces hasta la calcificación completa del periápice y *se tomó* este conjunto como patrón. (<www.dentalw.com>)

5.2. Characteristics of metadiscourse in popular science articles in Spanish

Phatic forms addressed to the reader / Colloquial phrases / Personalisations

Examples:

(4) *¿Sabía usted que* el invento de la pila eléctrica no tenía como objetivo la fabricación de un dispositivo de almacenamiento de energía?. *¿O que el proceso de vulcanización del caucho fue fruto de un accidente? ¿Cuántos premios Nobel han tenido su origen en un error de síntesis?*

Seguro que podemos adivinar los sarcásticos comentarios de los pragmáticos de la época. *¡Valiente negocio!,* emplear cuatro volúmenes de gases para generar electricidad que genera un solo volumen. *¡Menuda pérdida de tiempo!*

Otro punto muy flojo en la cadena de bioseguridad son los propios médicos, que *poseen una ignorancia total sobre el tema*, a diario *vemos* en los laboratorios *en que forma remiten las muestras* y mucho mas aun si ellos son cirujanos. (<www.cienciateca.com>)

5.3. Characteristics of metadiscourse in research articles in English

Hedges or modal forms / Expressions that limit conditions or knowledge

Examples:

(5) On a global basis, *approximately* 39.5 million people are infected by the HIV virus.

There are several reasons to be skeptical about the claim that nanotechnology *could be* of assistance to the developing world.

Lab-on-chip technology will usher in a new era that *could reshape* society in profound ways. (*Bulletin of Science, Technology and Society* <http://bst.sagepub.com>)

5.4. Characteristics of metadiscourse in popular science in English

Personalisations and direct verbal forms addressed to the reader / Phatic forms / Colloquial vocabulary

Examples:

(6) *You cannot plug your laptop into an electrical outlet or recharge your cell phone,* iPod or cordless tools without each item's custom DC adapter. *Wouldn't it be handy if one little black brick covered them all?*

The problem? The "brick" *would have to ask every gadget* what voltage and current to send. (<www.sciam.com>)

6. Discussion and conclusion

The general conclusions we can draw suggest that the main differences between English and Spanish metadiscourse in research articles and popular science are cross-generic more than intra-generic. In other words, metadiscourse has proved to be very similar in English and Spanish within the same genre, with the exception of very specific traits such as the passive voice in English or the impersonal forms in Spanish. However, deeper and wider research proves necessary in order to explore frequency of metadiscursive elements in both languages and genres. Also, the field of specialisation variable should be added, should this be medicine, physics, chemistry, etc. so as to check whether there exist differences within each field. In any case, the purpose of this study was not to ascertain the frequency of metadiscursive elements, but to raise awareness about the need to take into account metadiscourse in the translation of scientific texts in research articles and popular science, probably the two scientific genres which are translated the most nowadays, due mainly to electronic journals published on the Internet.

Not considering metadiscourse an important part of texts and therefore not translating it according to the target language prescription would lead to deficient translations that could impair the communicative goals of these two scientific genres, which, as has been said, are quite different. Translation of scientific texts, viewed from the discourse analysis perspective, not only has to consider the communicative goals and functions belonging to each genre as crucial aspects of it, as much as register characteristics are, but also metadiscourse, if we want to produce high quality translations.

References

Alcaraz Varó, Enrique 2000. *El inglés profesional y académico.* Madrid: Alianza Editorial.

Bédard, Claude 1986. *La traduction technique, principes et pratique.* Montréal: Linguatech.

Cabré, Maria Teresa 1993. *La terminología. Teoría, metodología, aplicaciones.* Barcelona: Antártida.

Crismore, Avon 1989. *Talking with Readers: Metadiscourse as Rhetorical Act.* New York: Peter Lang.

Crismore, Avon / Farnsworth, Rodney 1989. Mr. Darwin and his Readers: Exploring Interpersonal Metadiscourse as a Dimension of Ethos. *Rhetoric Review.* 8/1, 91-112.

Crismore, Avon / Farnsworth, Rodney 1990. Metadiscourse in Popular and Professional Science Discourse. In Nash, W. (ed.), *The Writing Scholar: Studies in Academic Discourse.* Sage: Newbury Park/London, 118-136.

Crismore, Avon / Markkanen, Raija / Steffensen, Margaret 1993. Metadiscourse in Persuasive Writing: A Study of Texts Written by American and Finnish University Students. *Written Communication.* 10/1, 39-71.

Dahl, Trine 2004. Textual Metadiscourse in Research Articles: A Marker of National Culture or of Academic Discipline? *Journal of Pragmatics.* 36, 1807-1825.

Fuertes-Olivera, Pedro Antonio / Velasco-Sacristán, Marisol / Arribas-Btio, Ana / Samaniego-Fernández, Enrique 2001. Persuasion and Advertising English: Metadiscourse in Slogans and Headlines. *Journal of Pragmatics.* 33, 1291-1307.

Gamero Pérez, Silvia 2001. *La traducción de textos técnicos.* Santa Perpètua de Mogoda, Barcelona: Ariel.

Halliday, Michael A.K. 1978. *Language as Social Semiotic: The Social Interpretation of Language and Meaning.* Baltimore: University Park Press.

Hyland, Ken 1998. Persuasion and Context: The Pragmatics of Academic Metadiscourse. *Journal of Pragmatics.* 30, 437-455.

Hyland, Ken 1999. Talking to Students: Metadiscourse in Introductory Coursebooks. *English for Specific Purposes*. 18/1, 3-26.
Hyland, Ken 2000. *Disciplinary Discourses: Social Interaction in Academic Writing*. London: Pearson.
Hyland, Ken / Tse, Polly 2004. Metadiscourse in Academic Writing: A Reappraisal. *Applied Linguistics*. 25/2, 156-177.
Maillot, Jean 1997. *La traducción científica y técnica*. Madrid: Gredos.
Mauranen, Anna 1993. Contrastive ESP Rhetoric: Metatext in Finnish-English economics texts. *English for Specific Purposes*. 12, 3-22.
Montalt, Vicent 2005. *Manual de traducció cientificotècnica*. Vic: Eumo Editorial.
Nord, Christiane 2000. What Do We Know about the Target-Text Receiver? In Beeby, A. / Ensinger, D. / Presas, M. (eds) *Investigating Translation*. Amsterdam: John Benjamins Publisher, 195-212.
Nord, Christiane 2007. The Phatic Function in Translation: Meta-communication as a Case in Point. In Vandeweghe, W. / Vandepitte, S. / Van de Velde, M. (eds) *The Study of Language and Translation*. Amsterdam: John Benjamins Publisher, 171-184.
Salager-Meyer, Françoise 2004. Hedges and Textual Communicative Function in Medical English Written Discourse. *English for Specific Purposes*. 13/2, 149-70.
Suau-Jiménez, Francisca 2001. El género y el registro en la traducción del discurso profesional: un enfoque funcional aplicable a cualquier lengua de especialidad. Universidad Autónoma de Barcelona. Barcelona: *Revista Electrónica CULTURELE*. <http://www.ub.es/filhis/culturele/tinasuau.html>
Suau-Jiménez, Francisca 2005a. Matizadores discursivos frente a elementos apelativos y fáticos en artículos científicos. In Gil-Salom, L. / Aguado de Cea, G. (eds) *Estudios sobre las lenguas de especialidad en España*. RESLA, Volumen Monográfico, 125-134.
Suau-Jiménez, Francisca 2005b. Metadiscourse in Research and Popular Science Articles or How to Please the Audience: A Cross-Generic and Intra-Generic Analysis in Search of a Common

Metadiscursive Core. *IPRA International Conference*. Riva del Garda, Italy, July 2005.

Suau-Jiménez, Francisca 2005c. The Difficult Balance between Author's and Scientific Community's Power over Research Articles in Applied Linguistics. *LSP and Professional Communication*. 5/2, 59-72.

Suau-Jiménez, Francisca Forthcoming. El Metadiscurso en el género "Servicios y Productos Turísticos" en inglés y español: importancia de su traducción como recurso para la persuasión del cliente. *I Congreso Internacional de Traducción Especializada*. Colegio de Traductores Públicos de la ciudad de Buenos Aires. Universidad de Buenos Aires: Argentina, July 2006.

Suau-Jiménez, Francisca / Dolón-Herrero, Rosana 2007. The Importance of Metadiscourse in the Genre 'Promotion of Touristic Services and Products': Differences in English and Spanish. In Galová, D. (ed.) *Languages for Specific Purposes: Searching for Common Solutions*. Cambridge: Cambridge Scholars Publishings, 70-79.

Suau-Jiménez, Francisca / Dolón-Herrero, Rosana Forthcoming. El metadiscurso en la traducción de textos performativos. *Congreso Internacional de AESLA*. Almería, April 2008.

Valero-Garcés, Carmen 1996. Contrastive ESP rhetoric: Metatext in Spanish-English economics texts. *English for Specific Purposes*. 15/4, 279-294.

Vande Kopple, William J. 1985. Discourse about Discourse. *College Composition and Communication*. 36, 82-93.

Olga Torres-Hostench / José Ramón Biau Gil / Pilar Cid Leal / Adrià Martin Mor / Bartolomé Mesa-Lao / Mariana Orozco / Pilar Sánchez-Gijón

TRACE: Measuring the Impact of CAT Tools on Translated Texts

1. Introduction

The information society and the new technologies it has generated have had an important impact on translation in general and on scientific and technical translation in particular.

With the introduction and widespread use of Computer-Assisted Translation (CAT) tools, in particular the use of translation memories, significant changes have been brought about in the translation workplace. Indeed, multilingual communication in the field of science and technology has become dependent upon computer-assisted translation, and even automatic translation, to such an extent that presentations and texts are often prepared in the original language and then simplified in order to facilitate translation using new technologies.

CAT tools have been incorporated into the translation process, generally speaking, for the following reasons: a) they can save a significant amount of time compared to translating without CAT tools; b) they facilitate managing projects that involve translating a source text into various languages; c) they allow previous translations to be re-used as a language resource for new translations; d) they simplify the handling of files in different formats and facilitate page layout or DTP and final publication of the translations; e) they systematise the translation process so that it can be standardised and protocols can be implemented; and f) they establish different professional roles within the translation process, and thus encourage specialisation among translation professionals.

Whilst there is little doubt that CAT tools are beneficial to companies and professional translators alike, little research has been carried out to determine the differences that may exist between technical texts translated with or without CAT tools, or the impact that the practice of translating using CAT tools may have on developments in target language usage. The aim of our research is to investigate the effects of the use of computer-assisted translation tools on target text production.

2. Research aims and hypotheses

The aim of the TRACE project (*Traducción Asistida, Calidad y Evaluación*[1]) is to carry out an experiment to compare translations made with and without the use of CAT tools.

Our starting point was the following construct: "Translations resulting from different processes are different". By reformulating this construct, we can focus on CAT tools as follows: "the use of CAT tools has an impact on the final translation". Working from this premise, our objective is to create an environment and the necessary instruments to be able to determine how the use of CAT tools affects the translation product.

Based on the aforementioned construct, our theoretical hypothesis is as follows: "the final text obtained through a translation process using CAT tools differs from a translation process in which CAT tools have not been used".

Three study hypotheses derive from this initial hypothesis:

i) The phenomenon of explicitation is expressed differently in translations made with or without CAT tools.
ii) The phenomenon of linguistic interference is expressed differently in translations made with or without CAT tools.

1 Project HUM-04349-FILO of the Spanish Ministry of Education and Science, 2006-2009.

iii) The phenomenon of textuality is expressed differently in translations made with or without CAT tools.

3. Experiment design

In order to test our hypotheses, it was decided to carry out an experiment using professional translators working in a simulated real-world environment.

3.1. Independent variable

One independent variable – the presence or absence of CAT tools during the translation process – was established. Three different scenarios were determined:

- Translation without using CAT tools. Translation is performed using MS Word, the most commonly used word processor in the translation field. This scenario is referred to as SCE1.
- Translation using CAT tools. This scenario has been sub-divided into:
    - Translation using CAT tools which provide a WYSIWYG interface. Translation is performed using a combination of Workbench by SDL Trados and MS Word – a common combination of tools with these screen display characteristics. This scenario is referred to as SCE2.
    - Translation using CAT tools without a WYSIWYG interface. Translation is performed using a combination of Workbench and TagEditor by SDL Trados – a very common combination of CAT tools with these particular screen display characteristics. When using TagEditor, subjects do not see the original format of the source document, only a plain text with no pictures, background colour, different fonts, etc. This scenario is referred to as SCE3.

3.2. Dependent variables

Three dependent variables were established – two of which are considered to be translation universals (explicitation and linguistic interference) and a third, the textuality of the translated texts. (It should be noted that the purpose behind the selection of these three variables is not to establish which translation is "better". Our objective is to establish how translations differ on the basis of a given set of parameters, according to whether CAT tools are used or not, in a comparative study of translations resulting from different translation processes.)

3.2.1. Explicitation

Explicitation was first described by Vinay/Darbelnet in 1958 as the "procédé qui consiste à introduire dans LA des précisions qui restent implicites dans LD, mais qui de dégagent du contexte ou de la situation" (1995: 9), which can be translated as "the process of introducing into the target language information details which remain implicit in the source language, but which emerge from the context or the situation". What then became known as the "explicitation hypothesis" was formulated by Blum-Kulka in 1986. In its historical development, this hypothesis broadly states that a translation will be more explicit than a corresponding non-translation, which may be either the source text or a parallel text in the target language.

Explicitation occurs, for example, when a ST (source text) unit of a more general meaning is replaced by a TT (target text) unit of a more specific meaning; the complex meaning of a ST word is distributed over several words in the TT; new meaningful elements appear in the TT, etc.

Blum-Kulka's explicitation hypothesis has generally been confirmed by studies on translation between different language pairs (Séguinot 1988; Weissbrod 1992; Klaudy 1993; Englund 1993; Øverås 1998; Olohan/Baker 2000; Whittaker 2004). However, this hypothesis has never been tested working with texts produced with and without computer-assisted translation tools. Hence, the importance of analysing this dependent variable in our research project.

The explicitation-related parameters used to obtain data under experimental conditions are as follows:

1. the proportion of explicit to implicit cultural referents;
2. the proportion of phoric to fully lexical (auto-semantic) phrases (pronouns);
3. the number of newly introduced cohesive referents per discourse segment;
4. the strength of lexical units in terms of hyperonimic to hyponimic relationships.

These four parameters are partly based on previous findings made by the corpus-based project CroCo from Saarland University (Germany)[2], and they aim to provide further data for explicitation research.

Further research in this area may prove useful to shed light on explicitation in translation as a possible translation universal. It may also prove useful to determine the extent to which the translation environment affects translators' output.

3.2.2. Linguistic interference

In 1995, Toury introduced the notion of linguistic interference as a translation universal – a notion first defined by Baker in 1993. In his classification in 1996, Baker makes reference to four universals: *explicitation, simplification, normalisation/conservatism* and *levelling out*. Toury adds *interference* as a universal of translation, as well as *increasing standardisation*. Regarding *interference,* Toury states that: "In translation, phenomena pertaining to the make-up of the source text tend to be transferred to the target text". (Toury 1995: 275)

Besides factors such as a translator's experience or the socio-cultural context of the language pair involved in a translation, some authors have investigated how the use of translation memories affect the degree of linguistic interference in a translation (Dragsted 2004; Bowker 2005). Some studies have even investigated this effect in relation to the quality of the translation product (Bédard 2000; Bowker 2005; Vilanova 2006), although the number of subjects

2 CROCO: *Sprachliche Eigenschaften von Übersetzungen* – eine korpusbasierte Untersuchung für das Sprachenpaar Englisch-Deutsch (DFG-Projekt) <http://fr46.uni-saarland.de/croco/index.html> last accessed: 15th April 2009.

involved in the studies conducted was small and they did not always involve professional translators.

The indicators used in our experiment to determine the effect of CAT tools on linguistic interference between Spanish and English were divided into eight different categories: typography and spelling; vocabulary; morphosyntax; text conventions; subject matter; encyclopedic knowledge; cohesion and coherence.

3.2.3. Textuality

The concept of textuality, adapted from Neubert (1992), includes the following three phenomena:

a) Target language rules: the source texts include punctuation items that are correct in the source language, but incorrect in the target language. The experiment will test whether the produced target texts apply target-language rules or stick to the source text punctuation.
b) Internal information coherence: each source text includes information items that can be interpreted in different ways in isolation, but that can only have one possible interpretation when analysed together with the rest of the verbal and non-verbal text elements. The experiment will test whether internal information coherence is kept in the target texts.
c) Target-reader perspective: each source text includes information items that do not belong to the target culture. The experiment will test whether, in the target texts, target-culture readers have been taken into account by adding or modifying such information items.

3.3. Subjects

Having decided to carry out an experiment using professional translators as subjects, the following selection criteria were established: a) subjects should be professional translators; b) with English-Spanish as one of their professional language pairs; and c) be habitual users of SDL Trados, with sufficient knowledge of the tools to translate a text with each of those tools with no external assistance.

Both experienced and novice freelancers replied to our call for a paid, one-off collaboration performing a translation task for the university. In-house translators were located by contacting translation and localisation companies and asking them to recruit some of their in-house professionals for a specific translation task. The in-house translators had at least two years of experience working in-house, most of them reaching or exceeding the three years that were required from experienced freelancers. For selection purposes, a questionnaire was administered to each applicant to obtain information concerning their professional background and their knowledge of the use of CAT tools.

Over 90 subjects were selected and divided into three groups: a) freelancers with more than three years of experience; b) in-house translators, most of them recommended and sent by their employers; and c) freelance translators with less than two years of professional experience.

Subjects whose answers to the questionnaire proved to be unreliable who required assistance when performing the requested tasks as a result of shortcomings in their knowledge of SDL Trados (usually TagEditor) or who had not completed all the tasks required (as evidenced in a post-translation questionnaire) were subsequently disregarded for the purposes of the study. By disregarding these subjects it was possible to control extraneous variables such as translation speed and computer literacy.

3.4. Data collecting instruments

Data-collecting instruments used in the final experiment included: i) texts with indicators of explicitation, linguistic interference and textuality; ii) translation memory; iii) recordings of all operations performed using the keyboard and mouse, in XML format; iv) video recordings of the screen activity; and v) a post-translation questionnaire.

3.4.1. Texts and indicators

Given that the objective of our experiment was to determine the effect of the use of CAT tools on texts translated in three different scenarios, the selection of texts and indicators that would provide the data necessary to test our hypotheses was of primary importance.

This was because the format or subject-matter of a specific text, which might prove to be ideal for one scenario, might not prove to be so appropriate for another. For example, translating software interface today using MS Word without using CAT tools might, on the one hand, preserve the product file format and, on the other, maintain the strict level of linguistic coherence required by the product, but it would certainly be an unlikely scenario in the professional workplace.

Thus, three texts were selected that were deemed appropriate for translating in work environments with or without CAT tools. All three texts dealt with some aspect of information technology and were similar in both length and content. Each included several paragraphs and headings, together with a picture to which some reference is made in the text itself. Once the texts had been selected, they were modified slightly to ensure that certain characteristics (length, level of difficulty, etc.) were comparable. None was longer than 500 words.

Some sections of the texts were also modified to include indicators of variables selected for the experiment. The texts were then evaluated by several external researchers and professional translators to ensure that they were well suited to our needs.

As the texts used in our experiment had to be comparable in length and difficulty and appropriate for use in the three different scenarios proposed, the indicators of our dependent variables had to be comparable across the three texts.

The first proposal for indicators was based on the assumptions made in the study hypothesis. To test the usefulness of these indicators, the texts in which they were embedded were sent to be translated by professional translators in a translation company. The three different scenarios were used, coinciding with those planned for the final experiment. The resulting translations allowed the validity of the initial indicators to be determined. Other indicators, which had not originally been anticipated and which proved to be highly informative, were then added and all indicators were subsequently submitted for evaluation by university lecturers in Translation in the form of a questionnaire with a list of the proposed indicators. Lecturers were invited to evaluate whether or not each of the indicators was an appropriate parameter for our dependent variables.

3.4.2. Translation memory

When undertaking technical translation projects involving the use of CAT tools, translators normally have at their disposal a previously established translation memory. The translation memory provided in our experiment, however, was empty for the following reasons:

- Previous translations could interfere in the presence or absence of phenomena related to explicitation or linguistic interference. Translators would base their translations on previous translations and make translation decisions which from the outset would restrict their approach to the text and work environment. Should this occur, it would be impossible to determine whether more or less cases of explicitation or interference were attributable to the work environment.
- The CAT tool translation environments would have more information available than the environment with no CAT tools. It is not the same to query information in a translation memory as to search for information in parallel texts.

3.4.3. Recordings

The time taken to translate each text and the resources used by translators during the translation process were recorded for triangulation with data obtained from other sources to evidence possibly related patterns. A record was kept of all the operations performed using the keyboard and mouse in XML format, as well as a recording in video format of all the operations performed on screen.

3.4.4. Questionnaires and session logs

A first questionnaire requesting personal data and other information was administered to subjects as part of the selection process. A second, evaluation questionnaire, was designed for lecturers' evaluation of indicators. A third, post-translation questionnaire, was administered after subjects participating in the experiment had completed their translations, to determine the level of their satisfaction with the translation product. Session logs recorded all incidents occurring in the computer room during the pilot study and the experiment.

The experiment protocol designed to manage the final experiment was trialled in a pilot study in which participants translated all three texts previously selected and validated in all three scenarios programmed.

4. Phases of the research project

The TRACE project has developed through the following phases:

1. Design of the experiment: theoretical hypotheses, study hypotheses, variables, and selection criteria for experimental groups.
2. Design of data-collecting instruments: texts for translation incorporating indicators of the three dependent variables established; an evaluation questionnaire to evaluate texts and indicators.
3. Exploratory test to validate the appropriateness of the texts and indicators selected, and to introduce any necessary changes in the indicators based on the feedback received.
4. The design of a questionnaire to select subjects for the experiment; a protocol to manage the final experiment; and a post-translation questionnaire.
5. Pilot study to trial the data collecting instruments (texts, questionnaires, recordings etc.) and the experiment protocol.
6. Data analysis – analysis of data obtained from the pilot test.
7. The experiment – a controlled experiment involving 72 professional translators working in three different scenarios within a simulated ‘real-world’ context.
8. Data analysis – analysis of data obtained from the experiment.
9. Conclusions.

A more detailed description follows of the exploratory test, pilot study and final experiment.

4.1. Exploratory test

Once our data-collecting instruments (texts and indicators, and evaluation questionnaire) had been designed, in June 2008 the final versions of each were submitted to experts for evaluation. Evaluation was carried out by seven English-Spanish translation professionals who were also university lecturers at the Universidad Autònoma de Madrid, Universidad de Granada, Universitat Pompeu Fabra (Barcelona) and Universitat Autònoma de Barcelona. The instrument used for evaluation purposes was a closed questionnaire. The aim of our exploratory test was to ensure that the indicators selected were indeed valid for our research and, at the same time, confirm that each of the texts selected was of a similar level of difficulty.

Lecturers were asked if, in their opinion, a specific translation problem could be measured by a specific indicator in its context, and were invited to group the indicators they were given into different categories. After reading the three texts submitted for evaluation, they were also invited to suggest new or alternative indicators. Finally, they classified each text as 'very easy', 'easy', 'average', 'difficult' or 'very difficult'.

4.2. Pilot study

Once the texts and indicators to be used in the experiment had been chosen and validated, the experiment protocol was trialled. To do this, 18 postgraduate students in the Master's degree courses in Translation Technologies and in Audiovisual Translation were invited to participate in a pilot study. Students from both courses had received extensive training in the use of SDL Trados, TagEditor, and Translators' Workbench during the previous months. Their profile was thus similar to that of professional translators (of particular interest in this study).

In order to control tool-order and text-order variables, a complete tool-order combination (all possible tool orders were tested) and a simplified text-order combination (all texts were translated in the three possible positions, but not all combinations were tested, since texts were not our independent variable) was designed. This provided

18 possible text and tool combinations. The number of participants in the pilot study (18) was determined by the minimum number required to work with all the six possible combinations of translation environments (SCE1+SCE2+SCE3) and a minimum of three possible text combinations (T1+T2+T3), i.e. 6x3.

Students translated all three texts selected in two different sessions (two in the first session, and a third text in the second session). This was because it was considered that the texts were perhaps too long for them, and we did not want translation speed or time pressure to affect results. All students used clone computers in the same computer room, with exactly the same hardware and software (the latter thanks to an image created specifically for the pilot study). As each translation was completed, it was sent by email to one of the researchers. When the translation was received, access was given to the next translation. This way, students were prevented from modifying their first text after seeing the second.

Several methods were used to collect data not only about the translation product, but also about the translation process:

a) Direct observation: Two researchers were always present at each session. Students were told that these two people were "technical support" in case there was a problem, but the researchers also kept a log of the whole room activity (who needed help, eventual technical problems, beginning and end of the session, etc.).
b) Recording software: Each computer was equipped with a screen-recording tool and a keyboard-logging tool to record everything that appeared on the screen or was typed on each computer. These videos and logs were used to ensure that each student had followed the instructions given for each text (e.g. type of software to be used for each translation, etc.). They were also used to measure the time needed to translate each of the texts.

4.2.1. Results of the pilot study

As a result of the pilot study, some changes were made to the way in which the final experiment was to be conducted. Firstly, it was observed that too much time had been allotted for the translation of each text.

The decision was taken to reduce the time allotted so that each text could be translated in a single session, thus reducing the presence of external variables. Secondly, some students complained about keyboard performance. It was discovered that only those using TagEditor were affected, and that the problem disappeared when InputLog, the keylogging program, was not activated. It was thus decided not to use the keyboard logger with TagEditor in the final experiment. Improvements were made in some organisational aspects of the experiment. The pilot study also allowed researchers both to confirm the validity of the indicators they were working with and to begin to contemplate more useful parameters for data analysis, to be exploited after conducting the final experiment.

4.2.2. Data analysis

Statistical analysis with SPSS was used to process data obtained from the pilot study and the following preliminary results were obtained for the indicators established:

Preliminay results on explicitation

The pilot study carried out did not appear to validate our hypothesis with regard to explicitation: there were no clear differences in explicitation for any of the three environments. Nor did statistical analysis of the different indicators for this dependent variable appear to evidence any significant pattern. There was more evidence of explicitation in SCE1 than in SCE3.

These results have led us to establish a new research hypothesis, since SCE2 (Workbench + MS Word) seemed to be the translation environment in which fewer cases of explicitation occurred in comparison with the other two scenarios SCE1 and SCE3. Could it be the case that hybrid translation environments (word processor with TM functionalities) interfere to a greater extent in a translator's output?

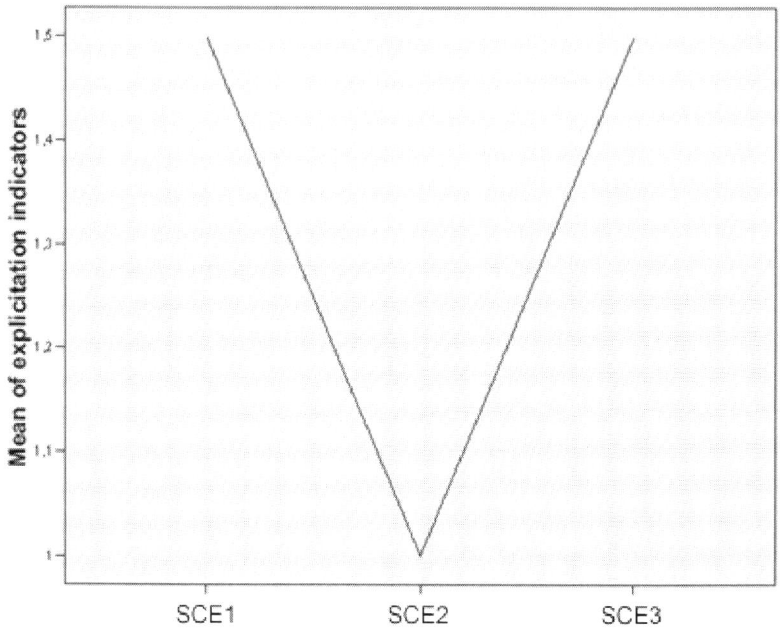

Graph 1. Mean of explicitation indicators in the pilot study.

Preliminary results on interference

Our initial hypothesis was only partially proved in the pilot study. Graph 2 shows how in scenario SCE 3 (Workbench + TagEditor) linguistic interference occurs in more than 50% of indicators analysed in fourteen out of sixteen subjects. In scenario SCE2, it occurs in more than 50% of indicators in a smaller number of subjects (11 out of 16), than in SC3, as we expected, but note that it also occurs in a slightly smaller number of subjects than in scenario SCE1 (12).

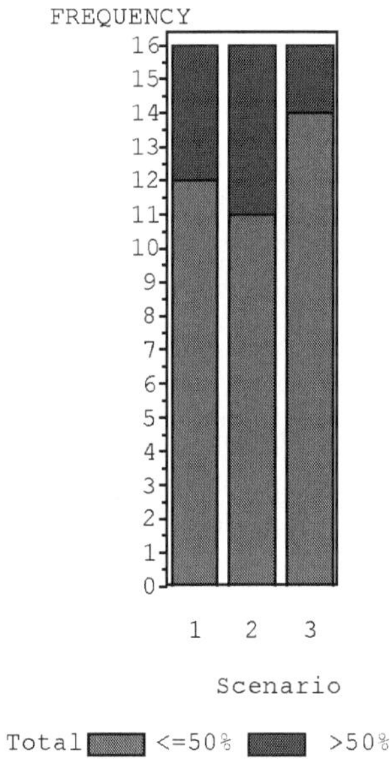

Graph 2. Linguistic interference in the pilot study.

Preliminary results on textuality

Data show that, when segmenting tools are used, there is a higher risk that intrasegmental faithfulness to the source text exceed (consciously or unconsciously) the intersegmental faithfulness to the target-language punctuation rules. That is, when subjects used SDL Trados, there were more violations of Spanish punctuation rules than when Microsoft Word was used. Further research was required in order to determine whether such violations are conscious (faithfulness to the source-text punctuation) or unconscious (unawareness of a target-language punctuation violation).

4.3. The experiment

To minimise the effect of external variables and to monitor the translation process more closely, the experiment was carried out in the Faculty of Translation and Interpreting of the Universitat Autònoma de Barcelona where the computer rooms could accommodate the large number of professional translators initially participating in the experiment. It was, however, decided to divide the study group into smaller groups to facilitate management of the experiment so that between January and March 2009, a total of twelve sessions were programmed on Friday mornings and afternoons.

Subjects who fulfilled the selection criteria for the experiment were invited to attend the Universitat Autònoma de Barcelona to perform a test translation for which they were paid. None was aware of exactly what the test involved. On arrival they were received in a welcome room and given instructions as to what they were required to do throughout the day. They were told that their translations formed part of a recruitment campaign for the Translation Services Office in the University and that their translations and contact details would be sent on to the Translation Service for future collaboration (our research group had already come to this agreement with the Translation Services Office). Subjects were then taken to the computer room, where the programs used to record keyboard, mouse and screen activity had been activated, before they gained access to the room. This was to ensure that subjects were unaware that their operations were being recorded. Each subject had then to translate the first text they had been given, in the environment or scenario they had been assigned in a printed personalised project brief. As soon as they had finished their translation, subjects returned to the welcome room to enjoy a coffee break, where three researchers engaged in small talk to minimise the possibility of subjects exchanging information about the texts they had just translated. Before subjects returned to the computer room to translate their second text, the three researchers who had played the role of "technical support" during the session backed up the target text of each first translation, plus the key-logging files and screen videos, and the recording software was then re-run for the next translation.

Table 1 shows the combinations and order in which texts and tools were used:

Tool order (6 different tool)	*Text combination* (3 different tool)
Tool order 1	Combination 1 (4 subjects)
	Combination 2 (4 subjects)
	Combination 3 (4 subjects)
Tool order 2	Combination 1 (4 subjects)
	Combination 2 (4 subjects)
	Combination 3 (4 subjects)
Tool order 3	Combination 1 (4 subjects)
	Combination 2 (4 subjects)
	Combination 3 (4 subjects)
Tool order 4	Combination 1 (4 subjects)
	Combination 2 (4 subjects)
	Combination 3 (4 subjects)
Tool order 5	Combination 1 (4 subjects)
	Combination 2 (4 subjects)
	Combination 3 (4 subjects)
Tool order 6	Combination 1 (4 subjects)
	Combination 2 (4 subjects)
	Combination 3 (4 subjects)

Table 1. Combinations of CAT tools and texts in the experiment.

After completing the third of a total of three translations, subjects were invited to return to the welcome room for the last time to complete a post-translation questionnaire stating the level of their satisfaction with the translations they had performed. They were also asked to give their formal consent to the use of the results of their translation as part of a study. Since there was no guarantee that all the subjects would give their consent, over 90 subjects were initially selected to take part in the experiment.

Each morning and afternoon session lasted five hours. If subjects requested extra time to complete their translations, their break was shorter, but all subjects were required to leave the computer-room for at least five minutes before beginning their next translation.

As in the pilot study, all the computers were equipped with exactly the same hardware and software (image system) in order to prevent computer performance becoming a variable. At least two researchers were always present in the room, playing the role of "technical support" staff, and observing the session.

Data obtained from session logs, video recordings, the selection questionnaire, and informal conversations during breaks helped identify some subjects who had provided unreliable personal data prior to selection; who did not have the required knowledge of CAT tools; who had not had time to complete their translations; or who had not followed the specifications for each translation. These subjects were disregarded for the purposes of our study.

Quantitative data was obtained exclusively from the indicators embedded in the translated texts.

At this point in time, researchers are in the process of analysing the data obtained.

5. Quality assurance

An important aspect of the TRACE research project is precisely the control exercised over multiple aspects of the research design and implementation. Strict control was exercised over the quality of the different phases of development of the project ranging from the selection of subjects and the design and validation of data-collecting instruments to the type of resources, infrastructure, and procedures used in the final experiment. In a controlled experiment such the one we have just described the results obtained may be considered to be highly reliable.

The control exercised over different aspects of this project has served to effectively determine: i) the coherence of the aims and objectives of the project and its viability; ii) the most appropriate research methodology to be used, the different phases of research and their distribution over time; iii) the research design; distribution of tasks and responsibilities within the group; and iv) the resources necessary to carry out the experiment. It was especially useful in the preparation and execution of the pilot study which evidenced both positive and negative aspects of the research design and allowed necessary adjustments to be made in preparation for the final experiment.

The detailed preparation and validation of data-collecting instruments in exploratory tests, the adjustments made to the translation environment as a result of the pilot study and the tight control of all activities involved in the experiment ensured that the final experiment was conducted successfully. All incidents were registered and carefully analysed so that they could be promptly resolved and the experiment carried out virtually according to plan. The experiment itself was documented in detail to facilitate data analysis and subsequent publication of the results obtained.

6. Conclusions

This chapter has described the design and development of the TRACE research project which is investigating the impact of CAT tools on translated texts. Given that CAT tools are extensively used for the translation of scientific and technical texts, it was decided to carry out our experiment focusing specifically on the translation of technical texts.

As an ambitious research project with a sound methodological basis, the TRACE project is one of the few research projects in the field of translation that has been carried out using a large group of professional translators as subjects, in such strictly controlled conditions, and with a database containing thousands of entries.

We believe that the results of this research project will make a useful contribution to the field of translation, both at a theoretical and a practical level: from a theoretical point of view, by providing a solid basis for the academic study of translation based on experimental research; from the practical application point of view, the comparative study of translations performed with or without CAT tools will be useful to evaluate how translated texts are affected by CAT tools.

The TRACE project will not only help us to determine the impact that new tools have on the translation of scientific and technical texts; it also showcases the changes and the challenges that translators have to face in an environment in constant evolution.

Finally, just as language in the fields of science and technology evolves over time with the introduction of new terms to describe discoveries and new inventions, so language evolves as a result of the introduction of new technologies used in the multilingual translation or localisation of terms. Mobile phones have revolutionised the use of language in SMSs. Research such as that carried out in the TRACE Project shows the influence of another kind of technology – CAT tools – on language in translated technical texts.

We hope too that we can provide the community of researchers in the field of translation with data, results and conclusions that will serve to move our discipline forward so that it is not left behind by the world of new technologies.

References

Bédard, Claude 2000. Mémoire de traduction cherche traducteur de phrases. *Traduire*. 186, 41-49.

Baker, Mona 1993. Corpus Linguistics and Translation Studies: Implications and Applications. In Baker, Mona *et al.* (eds) *Text and Technology. In Honour of John Sinclair*, Amsterdam/Philadelphia: Benjamins, 233-250.

Baker, Mona 1996. Corpus-based Translation Studies: The Challenges that Lie Ahead. In Somers, Harold (ed.) *Terminology, LSP and Translation. Studies in Language Engineering in Honour of Juan C. Sager*. Amsterdam/Philadelphia: Benjamins, 175-186.

Bowker, Lynne 2005. Productivity vs. Quality? A Pilot Study on the Impact of Translation Memory Systems. *Localisation Focus*. 4/1, 13-20.

Blum-Kulka, Shoshana 1986. Shifts of Cohesion and Coherence in Translation. In House, Juliane *et al.* (eds) *Interlingual and Intercultural Communication. Discourse and Cognition in Translation and Second Language Acquisition Studies*. Tübingen: Gunter Narr, 17-35.

Dragsted, Barbara 2004. *Segmentation in Translation and Translation Memory Systems: An Empirical Investigation Of Cognitive Segmentation And Effects Of Integrating A TM System Into Thetranslation Process*. Copenhagen: Samfundslitteratur.

Englund, Birgitta 1993. Semantic Change in Translation: A Cognitive Perspective. In Gambier, Yves / Tommola, Jorma (eds) *Translation and Knowledge*. Turku: Centre for Translation and Interpreting, University of Turku, 285-296.

Klaudy, Kinga 1993. On Explicitation Hypothesis. In Klaudy, Kinga / Kohn, János (eds) *Transferre necesse est... Current Issues of Translation Theory*. Szombathely: Daniel Berzsenyi College, 69-77.

Neubert, Albrecht / Shreve, Gregory 1992. *Translation as Text*. Kent/ London: Kent State University Press.

Olohan, Maeve / Baker, Mona 2000. Reporting *That* in Translated English: Evidence for Subconscious Processes of Explicitation? *Across Languages and Cultures*. 1/2, 141-158.

Øverås, Linn 1998. In Search of the Third Code: An Investigation of Norms in Literary Translation. *Meta*. 43/4, 571-588.

Séguinot, Candace 1988. Pragmatics and the Explicitation Hypothesis. *TTR: Traduction, Terminologie, Rédaction*. 1/2, 106-114.

Toury, Gideon 1995. *Descriptive Translation Studies and Beyond*. Amsterdam/Philadelphia: John Benjamins.

Vilanova, Silvia 2006. L'impacte de les tecnologies de traducció en el català tècnic. <http://isg.urv.es/publicity/doctorate/tribunal_unic_2006/ Filanova_research_design_v021.pdf>

Vinay, Jean-Paul / Darbelnet, Jean 1958. *Stylistique comparée du francais et de l'anglais: Méthode de Traduction*. Paris: Didier.

Weissbrod, Rachel 1992. Explicitation in Translations of Prose-Fiction from English to Hebrew as a Function of Norms. *Multilingua*. 11/2, 153-71.

Whittaker, Sunniva 2004. Étude contrastive des syntagmes nominaux demonstratives dans des textes traduits du français en norvégien et des textes sources norvégiens: stratégie de traduction ou translationese? *Forum*. 2/2, 221-240.

Juan José Martínez-Sierra

Science and Technology on the Screen: The Translation of Documentaries[1]

1. Introduction

In the full age of image, scientific and technical texts have not remained detached from it, have broken the limitations that paper meant, and have strengthened their presence in the different audiovisual media that have been developing during the course of the last decades. Such is the case of those texts that are delivered to us via television or the Internet, two means of communication that make it possible for specialised texts to reach a quantitatively higher audience.

Obviously, translation is also present in those means. In this paper, we focus on television, where specialised texts take the form of documentaries. In these documentaries the word coexists with the image to communicate jointly the ostensible message. This combination of word and image generates a type of text with several characteristics that differentiate it from oral and written texts: the audiovisual text. Audiovisual translation is the appropriate general translation mode to transfer this kind of texts and, more specifically, to take care of their specialised translation.

Thus, firstly, in this chapter the main characteristics of the documentary genre are introduced. Then, the way is described in which scientific and technical programmes created abroad and in a foreign language can be made accessible thanks to translation modes such as voice-over or dubbing. Assuming that it is equally important to know about the nature of the texts at stake and about that of the means by which

1 This research was conducted as part of a project founded by the Spanish *Ministerio de Ciencia e Innovación (Programa José Castillejo).*

they are transmitted and the way they do so, a review of these translation modes is also made. Their role in many countries is emphasised – together with that of subtitling – as the most frequent modes for the translation and later emission of documentaries. Finally, the importance of appropriate training of future translators is mentioned.

The content of this chapter concerns mainly the Spanish case, although it could be possible to extrapolate it to the situation in other countries that also make use of modes such as voice-over for the emission of scientific and technical documentaries originally created in another language. On the other side, documentaries are just one of the possible formats used by scientific and technical texts for their audiovisual transmission (we could mention certain advertising videos, corporate videos, audiovisual training material, instruction manuals on video, and others of the like). Nevertheless, this chapter focuses on documentaries, given their greater impact on the general audience.

2. The documentary: definition and description

According to the Houghton Mifflin Dictionary, a documentary can be defined as a "work, such as a film or television program, presenting political, social, or historical subject matter in a factual and informative manner and often consisting of actual news films or interviews accompanied by narration" (2009: online). In other words, documentaries are informative television programmes "meant to 'teach' something about [a source] culture to the receptor culture" (Franco 2001a: 178). In a nutshell, it is an ideal programme format for the transmission of scientific and technical contents.[2]

Media web site (Spanish *Ministerio de Educación y Ciencia*) offers interesting data regarding this type of programme (2009: online):

2 A general conception of *documentary* is assumed in this paper, though we are aware that different subtypes can be identified: docudrama, docufiction, mockumentary, ethnographic film, mondo film, or political cinema, among others.

- Documentaries were born with the shooting of the first film ever in the history of cinema: *La sortie des usines de Lumière* (1895).[3] However, they have found their main market in the current television industry, in such a way that both in generalist television stations and in thematic television channels documentaries constitute the star programme format as far as the divulging of the aspects of past and contemporary cultures is concerned.
- There is a type of documentary for each scientific or artistic discipline: historical, cultural, social, political, about scientific research, or about animals, to mention a few. Certain documentaries make use of image or sound archives. Some include interviews. Some even resort to computer reconstructions, while others turn to dramatisation. All in all, documentaries display a heterogeneous nature that has brought about the debate of whether or not they can be labelled as an audiovisual *genre*, an issue that is dealt with in the next section.
- Concerning their production, it is laborious and requires months of work and, on many occasions, significant amounts of money. The main – and challenging – objective is to offer to a wide and varied audience a product which is appealing as well as clear in its exposition, at the same time complying with the entertainment component that every television programme should have.
- Television stations are usually the ones that produce documentaries. For example, the main European public televisions produce hundreds of hours of documentary programmes every year.[4] Co-productions also make their way, and agreements between televisions and international organisations and NPOs are common. A third option involves the purchase of documentaries from independent specialised production companies.
- It is a type of audiovisual product that is on top form these

3 Follow this link for further information about the history of documentaries: http://www.britannica.com/EBchecked/topic/167498/documentary-film.

4 For instance, RTVE (the national Spanish public television) broadcasts several documentary programmes on a weekly basis, such as *En portada*, *Crónicas*, *Documentos TV*, *Redes*, or *Grandes Documentales*, among others.

days.[5] Some productions are even premiered in cinemas or shown on IMAX screens.[6] The modern computer techniques, the miniaturisation of cameras, the advent of many thematic channels specialised in broadcasting documentaries, and the relatively new media format (such as CDs, DVDs, or even the Internet) have certainly contributed to this revival.[7]

- As a proof of the success of this type of programme it will suffice to mention some of the most important thematic channels specialised in documentaries, such as the American *Discovery Channel* and the winner of over 100 Emmy awards *National Geographic Channel* or the French *Odyssée* and *Planète*. We can also refer to the Spanish channels *Odisea*, *Natura*, or *Docu TVE*.[8]

3. The documentary as an audiovisual genre

In the previous section we echoed the question of whether documentaries constitute an audiovisual genre. We find the answer in

5 In this paper, we refer to documentaries, films, series, etc. as *products*. As Martínez-Sierra explains, "it is [...] possible to talk of consumption within the scope of audiovisual translation. In this case, the products are the films, series, documentaries, and the like that reach our screens" (2008: online).

6 For example, 23 documentaries were premiered in Madrid in the first half of 2008 (Equipo Reseña 2008).

7 In Spain, for example, the number of documentary programmes increased significantly – it practically doubled – in the 2007-2008 season. Albeit still being a genre of limited consumption, its viewers represented 3.1% of the total audience (Corporación Multimedia 2008: online). For further details, see the Appendix.

8 Documentary film festivals or film festivals including documentaries are also gaining ground. In Europe, for example, we can mention the Biberach Independent Film Festival, the Copenhagen International Documentary Festival, the Helsinki Documentary Film Festival (DocPoint), the International Documentary Film Festival of Navarra, or the Thessaloniki Documentary Festival, among others. Similarly, documentaries are included as a category in influential ceremonies such as the well-known American Academy Awards (the Oscars), the Spanish Goyas, the German Berlinale, or the European Film Awards.

Agost (1999: 80-93), who proposes the following classification of audiovisual genres:

- Dramatic genre:
  - Films (the audiovisual text par excellence) and telefilms, in which those aspects related to the script and to the relationship between the word and the image are crucial.
  - Series, the dramatic subgenre that has the biggest presence on television.
  - Cartoons, which are habitually dubbed in all countries, given the characteristics of the audience they address (typically children).
- Advertising genre. Texts belonging to this category are very much determined by economic, social, and political factors.
- Entertainment genre. These texts are highly heterogeneous: children's and teenagers' magazines, certain computer games, or some humorous shows.
- Informative genre. Documentaries stand out among the programmes belonging to this genre.

In this classification we can see that, despite not being classified as a genre on their own, documentaries do appear as belonging to a genre: the informative. Without wishing to be controversial, this conception is assumed here.

4. The translation of documentaries

As already pointed out, audiovisual translation makes it possible for documentaries originally produced in another language to cross the language barrier and to be screened in Spain. The documentary, just as any other audiovisual product, presents a combination of words, images, and sounds that differentiates it from written products – which are translated[9] – and oral products – which are interpreted. Thus, the

9 In the sense of *written* translation.

audiovisual text is a type of text different to the written text or to the oral text, and therefore requires suitable translation modalities. Let us see now what we can understand by *translation modality* and then consider the modalities that are generally used to translate documentaries. Some implications for the translator will follow.

4.1. Definition of modality

Hurtado's work is useful to define what we can understand by *translation modality*. Referring to translation in general, this author (1996: 370) mentions the *translation mode* as one of the elements (together with the source text's medium – written, oral, audiovisual, musical, etc. – and mode – the language variation depending on that medium) to categorise interlingual translation. Thus, according to Hurtado, the *translation mode* refers to the variation in the translation depending on the characteristics of the source and target texts' media (2001: 639), and it can be (1996: 374):

- Simple: if the same medium is maintained in the target text.
- Complex: if there are changes.
- Simple subordinated or complex subordinated: together with the characteristics of the simple and complex translation modes, there is a mixture of media in the source text. The translation is then conditioned by (or *subordinated to*) the constraints imposed by the different media.

Hence, the way in which translations are done determines the different translation modalities. Hurtado proposes the following classification, the categories of which are open to combination (1996: 374):

- Simple. Written translation (from written source text – ST – to written target text – TT), simultaneous interpretation (from oral ST to oral TT), liaison interpretation (from oral ST to oral TT), and *chuchotage* (from oral ST to oral TT).
- Complex. Sight translation (from written ST to oral TT) and consecutive interpretation (from written ST to oral TT).

- Simple subordinated. Theatre translation (from oral ST to oral TT), dubbing (from oral ST to oral TT), comic translation (from written ST to written TT), translation of advertising signs (from written ST to written TT), translation of word puzzles and crosswords (from written ST to written TT), and translation of songs (from oral ST to oral TT).
- Complex subordinated. Subtitling (from oral ST to written TT) and opera subtitling (from oral ST to written TT).

Hurtado does not mention it, but it seems possible to include the modality of voice-over within the *simple subordinated mode* category.

We can consider audiovisual translation a *general* translation modality – at the same level as written translation and interpretation –, which in turn can be divided into a series of audiovisual translation *specific* – not in the sense of *specialised* – modalities, with which we deal next.

4.2. Modalities for the audiovisual translation of documentaries

There are different audiovisual translation specific modalities, such as dubbing, subtitling, voice-over, simultaneous interpretation, narration, partial dubbing, free commentary, or sight translation. In this chapter, we focus on dubbing, voice-over, and subtitling, since they seem to be the most commonly used for the translation of documentaries (more marginally in the case of subtitling), as Agost explains (1999: 87-88):

- In the case of documentaries with off-screen narration in the source version – that is, when we can hear a voice but we cannot see the person to whom it belongs –, it is common to use off-screen dubbing or voice-over.
- To translate those documentaries in which the narrator – often a celebrity – can be seen, voice-over is usually employed, although dubbing is also used occasionally.
- In those documentaries including testimonies or even conversations, again both dubbing and voice-over are used. If several languages appear in the source version, it is possible to combine these two modalities with subtitling.

Let us see these three modalities in greater detail:

Modality 1

As Agost explains, *dubbing* consists in the substitution of the original soundtrack for another. Such substitution entails a change of language (otherwise we would be talking of *postsynchronisation*), and it must observe the following synchronisms (1999: 16):

- Content synchronism, which implies that the plot of the source and dubbed versions must be congruent with each other. Authors such as Chaume (2003a: 261) think that, in this case, the use of the term *synchronism* can be misleading and that we should talk of *coherence* instead, which escapes the limits of adaptation.
- Visual synchronism, which implies that the articulatory – and visible – movements of the mouth and the sounds we can hear are in concord.
- Characterisation synchronism, which implies harmony between the voice of the dubbing actors and actresses and the looks and gesticulation of the on-screen actors and actresses.

Chaume focuses on visual synchronism. For him, synchronisation "is one of the basic characteristics of translation for dubbing" and "consists of matching the TL translation with the articulatory and body movements of the screen actors, as well as matching the utterances and pauses in the translation with those of the ST". He also mentions that good synchronisation is crucial in dubbing, since it conveys "what Metz calls 'impression of reality', a powerful phenomenon basically consisting of the conjunction of images in motion, credible photography and icons, and credible dialogue and sounds" (2008: 129, 131).

Chaume distinguishes three types of *visual* synchronisation (2008: 132-136):[10]

- Lip synchrony: to "adapt the translation to the articulatory movements of the on-screen characters, making sure that the

10 See also Chaume (2004).

translation particularly respects the open vowels and bilabial and labio-dental consonants pronounced on screen".

- Isochrony: "To fit a (rough) translation into the duration of the screen characters' utterances, following their mouth articulation movements and their pauses and silences".
- Kinesic sinchrony: "to be coherent with the actor's body movements".

Whitman refers to a "myriad of intricate and interrelated 'synchronies'" that "must be fulfilled in order to produce high-quality target language versions" (1992: 19). Thus, in addition to *visual* synchronisation, she refers to *acoustic* synchronisation and discriminates between (1992: 39-53):[11]

- Idiosyncratic vocal type or "the compatibility between the voice of the acoustic personifier and the visual image of the actor on screen, his personality, character, deportment, etc". It could be argued that this feature will be culture bound, since each culture has its own catalogue of stereotyped voices.
- Paralinguistic and prosodic elements, or the consideration of "what the speaker does with his voice, the way he modulates it in terms of intensity, volume, speed, etc., to provide the hearer with information which surpasses the semantic content of the utterance".
- Cultural variations, accents and dialects, all of them carrying "undeniable, intricate 'messages', interpreted by the film-going audience with surprising consistency".

According to Whitman (1992: 54), "Much has been written about the hierarchy of synchronies," but "No agreement has been reached". Still, "most dubbing professionals [...] will concede that exacting and meticulous lip synchrony is an animal threatened with extinction". In this sense, Chaume (2008: 136) explains that, "in real professional practice, lip-sync is only observed in close-ups and extreme or big close-ups".

11 See also Agost (1999).

In any case, the degree to which these synchronisms are to be observed is relative, and it will depend on the media (cinema, television, etc.) and on the audience's customs (Agost 1999: 17). Dubbing is the preferred option in countries such as Spain, Italy, Germany, France, Great Britain (occasionally), Austria, Thailand, Japan, Switzerland, or Ireland.

Modality 2

According to Díaz-Cintas/Orero (2005: 473), *voice-over* is defined as follows:

> a technique in which a voice offering a translation in a given language [...] is heard simultaneously on top of the source language [...] voice. As far as the soundtrack of the original is concerned, the volume is reduced to a low level that can still be heard in the background whilst the translation is being read. It is common practice to allow the viewer to hear the original speech in the foreign language for a few seconds at the onset of the speech and to reduce subsequently the volume of the original so that the translated speech can be inserted. The translation usually finishes several seconds before the foreign language speech does, the sound of the original is raised again to normal volume and the viewer can hear once more the original speech.[12]

It is possible to find certain shared characteristics between voice-over and dubbing. However, in voice-over visual synchronisation is not as strict as in dubbing, as no attempt is made to hide the foreign origin of the programme, which is obvious.[13] Still, authors such as Orero talk about *voice-over synchrony*, which refers to the "ability to fit the translated message on top of the original speech, leaving a few seconds

12 Authors such as Matamala (2008: 117) refer to the confusion shrouding the term *voice-over*. In any case, for the purposes of the present piece of research the definition provided above may suffice.

13 As Franco (2001a: 178) explains, "To create the illusion of being *an* original is actually impossible in voice-over versions due to the simultaneous presence of *the* original. The illusion created by these versions is rather that of authenticity, of faithful reproduction of documentary testimonies".

at the beginning and at the end" (Matamala 2008: 118).[14] In Spain, as already mentioned, voice-over is a frequent modality for the translation and later screening of documentaries. Yet the usefulness of this modality is not restricted to this programme format. As Matamala explains, "it is also used to revoice non-fictional programmes in most countries over the world," such as interviews or current affairs programmes. Moreover, "it is used to revoice fictional TV programmes in many Eastern European countries, as well as some children's programmes in Scandinavia" (2008: 113, 118).

Modality 3

According to Agost (1999: 17-18), *subtitling* consists in the incorporation of written target language subtitles on the screen in which the source version of a film is being shown, in such a way that these subtitles coincide approximately with the lines of the screen actors and actresses. Its main difficulty is the necessary synthesis work on what is said on screen. Isochrony, stricter in cinema, also affects subtitling. Besides, the subtitles are subordinated to the reading speed of viewers. Also, the translator has to try to make the subtitles reflect the particularities of oral language, since it is a mode of discourse that is written to be read. Subtitling is the most common practice in countries such as Holland, Belgium, Denmark, Norway, Sweden, Finland, Portugal, Cyprus, and Greece, as well as in most South American countries.

As Chaves explains, the different audiovisual modalities constitute translation processes that do not necessarily apply to a whole film but to parts of it. Thus, we can find a combination of modalities within the same film (2000: 59-60). This is completely applicable to the case of documentaries, and it is in fact common for a documentary to combine, for example, dubbing (for the off-screen narration) and voice-over (for the different testimonies that may appear).

14 See Orero (2006).

4.3. Implications for the translator

We have already seen that for authors such as Agost the documentary is an audiovisual product that belongs to the informative genre. Similarly, González-Requena (1989: 39) considers that documentaries constitute an informative type of discourse. Hence, whether we talk of an informative *genre* or *discourse*, it seems clear that it is a type of audiovisual product with some features that differentiate it from other products such as films or television series, for example.

Chaume (2003b: 194-198) discusses some aspects that, given the aforementioned singularity of documentaries, are of clear interest for the translator:

- Documentaries are texts that show a verbal density that is proportionally greater to that of other audiovisual products such as films or television series. This greater density means a higher number of words per minute.
- There is also a higher degree of referentiality, in the sense that the linguistic sign is often accompanied by the image of the object to which it refers. It is therefore convenient and useful for the translator to watch the images as he or she does the translation.
- Unlike scientific and technical research articles published in specialised journals and addressed to experts, documentaries, as already pointed out, target a quantitatively higher and more heterogeneous public. They aim to spread knowledge among an audience who, *a priori*, has no specific or deep information about the subject matter in question. Thus, an informative and didactic purpose predominates, which calls for as many explanations as necessary.[15] The translator faces a rather simple textual configuration and an equally simple syntax, the main hurdle probably being the specialised lexis that the text may contain.
- Documentaries usually portray a series of predictible elements, which appear in an equally predictible order: different facts are

15 Franco (2001a: 178) refers to the procedures adopted by translators "aimed to explain, clarify, make the meaning of the cultural referent less vague to the target audience".

exposed in a descriptive and chronological manner, and the narration of those facts goes from the general to the specific.

- Regarding the constraints that the translator has to face, these would be typical of any audiovisual text, but of course with certain particularities. More concretely, the translator of documentaries does not have much room for *creativity*. As far as synchronisation is concerned, documentaries are recorded programmes that were filmed on location and subsequently synchronised in a recording studio. Moreover, a good deal of what we can hear is usually off-screen, which greatly facilitates the adaptation stage.
- Unlike the news, for example, documentaries are fairly homogeneous both from the thematic and from the discourse's interpersonal tenor points of view.

5. Voice-over training

Much has been written about dubbing and subtitling. However, as Matamala states, "Voice-over is one of the less known modalities of audiovisual translation," although it is also true that more research has been conducted in recent times (2008: 115).[16] This is why, in this section, we focus briefly on some didactic aspects concerning this mode.[17] As we already saw, voice-over can be used not just to revoice documentaries, but also to revoice other types of programmes, a fact that means a significant market for which new professionals need to be trained. However, "despite the boom of courses on AVT […], hardly any of them prepare students for voice-over" (Matamala 2008: 115).[18]

Matamala (2008: 118) highlights some of the issues at stake in the training of future voice-over professionals. The already mentioned

16 See Franco (2000a, 2000b, 2001a, 2001b) or Orero (2004, 2009), for example.

17 For didactic content regarding dubbing and subtitling see, for example, Díaz-Cintas (2008).

18 See also Díaz-Cintas/Orero (2003).

voice-over synchrony results in a target text that "tends to be more synthetic than the original". Thus, "students should [...] be trained to condense information using synonyms or alternative structures in which hesitations, fluffs and certain oral features without referential content are omitted". Besides, "translators must be able to render the translation of an oral product in an adequately reformulated written form," a requirement that also involves lots of practice for its fulfilment.

Matamala's work is also of interest because she identifies the different "challenges for the voice-over professional – and therefore the skills to be developed by students," namely (2008: 119):[19]

- Comprehension problems
- Content errors in the original
- Terminology and documentation processes
- Speakers, registers, and audiovisual transfer modes

6. Concluding remarks

Scientific and technical texts find in documentaries the perfect programme format for television, in such a way that they do not just reach experts but the general audience. Audiovisual translation has the appropriate specific modalities to make it possible for a documentary that was originally created in a given language to cross the language barrier and to spread knowledge – as well as to entertain – in countries in which another language is spoken. Among the different audiovisual translation modalities, dubbing and voice-over (and more marginally subtitling) are usually the chosen ones for the broadcasting of documentaries translated for the Spanish audience. Just as happens with any other audiovisual product, documentaries are audiovisual texts that present a series of characteristics that distinguish them from written or spoken texts. Besides, they are texts that, to a certain extent, can display some specialised terminology. Thus, the audiovisual translator must face their translation taking into account

19 See Matamala (2008: 119-121) for further details.

a series of aspects that differentiate the translation of documentaries from the translation of other audiovisual products such as films. Of course, appropriate training seems convenient so that future professionals can succeed in such a task.

References

Agost, Rosa 1999. *Traducción y doblaje: palabras, voces e imágenes*. Barcelona: Ariel.

Chaume, Frederic 2003a. El ajuste en la traducción para el doblaje. In García-Peinado, M. A. / Ortega-Arjonilla, E. (eds) *Panorama actual de la investigación en traducción e interpretación,* vol. 2. Granada: Atrio, 253-268.

Chaume, Frederic 2003b. *Doblatge i subtitulació per a la TV*. Vic: Eumo.

Chaume, Frederic 2004. Synchronisation in Dubbing: A Translational Approach. In Orero, P. (ed.) *Topics in Audiovisual Translation*. Amsterdam/Philadelphia: John Benjamins, 35-52.

Chaume, Frederic 2008. Teaching Synchronisation in a Dubbing Course: Some Didactic Proposals. In Díaz-Cintas, J. (ed.) The Didactics of Audiovisual Translation. Amsterdam/Philadelphia: John Benjamins, 129-140.

Chaves, María José 2000. *La traducción cinematográfica. El doblaje*. Huelva: Universidad de Huelva.

Corporación Multimedia <http://corporacionmultimedia.es/_menu_principal.html>. 23 April 2009.

Díaz-Cintas, Jorge (ed.) 2008. *The Didactics of Audiovisual Translation*. Amsterdam/Philadelphia: John Benjamins.

Díaz-Cintas, Jorge / Orero, Pilar 2003. Course Profile: Postgraduate Courses in Audiovisual Translation. *The Translator*. 9/2, 371-388.

Díaz-Cintas, Jorge / Orero, Pilar 2005. Voice-over. In Brown, K. (ed.) *Encyclopedia of Language and Linguistics (2nd edition)*. Oxford: Elsevier, 477-480.

Encyclopaedia Britannica <http://www.britannica.com/EBchecked/_topic/167498/documentary-film>. 20 April 2009.

Equipo Reseña 2008. *Cine para leer enero-junio 2008*. Bilbao: Ediciones Mensajero.
Franco, Eliana 2000a. *Revoicing the Alien in Documentaries. Cultural Agency, Norms and the Translation of Audiovisual Reality*. Unpublished PhD thesis. Leuven: Catholic University of Leuven.
Franco, Eliana 2000b. Documentary Film Translation: A Specific Practice? In Chesterman, A. *et al.* (eds) *Translation in Context, Selected Papers from the EST Congress, Granada 1998*. Amsterdam/Philadelphia: John Benjamins, 233-242.
Franco, Eliana 2001a. Inevitable Exoticism: The Translation of Culture-Specific Items in Documentaries. In Chaume, F. / Agost, R. (eds) *La traducción en los medios audiovisuales*. Castellón: Universitat Jaume I, 177-181.
Franco, Eliana 2001b. Voiced-over Television Documentaries. Terminological and Conceptual Issues for Their Research. *Target*. 13/2, 289-304.
González-Requena, Jesús 1989. *El espectáculo informativo*. Madrid: Akal.
Houghton Mifflin <http://www.houghtonmifflinbooks.com/eref/index.jsp>. 20 April 2009.
Hurtado, Amparo 1996. La traduction: classification et éléments d'analyse. *Meta*. 41/3, 366-377.
Hurtado, Amparo 2001. *Traducción y traductología*. Madrid: Cátedra.
Lumière, Louis 1895. *La sortie des usines de Lumière* (film).
Martínez-Sierra, Juan José 2008. The Contact between Cultures and the Role of Translation and the Mass Media. *Translation Journal*. 12/1.
Matamala, Anna 2008. Teaching Voice-over. A practical Approach. In Díaz-Cintas, J. (ed.) *The Didactics of Audiovisual Translation*. Amsterdam/Philadelphia: John Benjamins, 115-127.
Media Ministerio de Educación y Ciencia <http://recursos.cnice.mec.es/media/television/bolque7/pag8.htm>. 20 April 2009.
Orero, Pilar 2004. The Pretended Easiness of Voice-over Translation of TV Interviews. *JoSTrans*. 2, 76-96.
Orero, Pilar 2006. Synchronisation in Voice-over. In Bravo, J. M. (ed.) *A New Spectrum of Translation Studies*. Valladolid: Universidad de Valladolid, 255-264.

Orero, Pilar 2009. Voice-over in Audiovisual Translation. In Anderman, G. / Díaz-Cintas, J. (eds) *Audiovisual Translation. Language Transfer on Screen*. Basingstoke: Palgrave Macmillan, 130-139.
Whitman, Candace 1992. *Through the Dubbing Glass*. New York: Peter Lang.

Appendix

	2006-2007 Season	*2007-2008 Season*
News	16.2	17.6
Magazines	11.8	12
Foreign Series	11.9	11.1
Documentaries and Popularising Programmes	5	9.4
Films and Telefilms	8.9	8.7
Children's Programmes	6	6.8
Sports	4.3	6.5
Game Shows	12.6	6.1
Spanish Series	3.7	3.7
Shows	2.6	2.4
Reality Shows	1.7	1.8
Music Shows	0.7	1.8

Table 1. Percentage of broadcast time dedicated by the Spanish generalist television stations to the different types of programmes (seasons 2006-2007 and 2007-2008).

María Rosario Bautista Zambrana

Ontologies for Scientific-Technical Translation[1]

1. Ontologies and translation: basic concepts

Scientific-technical translation aims at transferring precise, univocal, clear and objective information from one language into another. It involves transferring not only certain syntactic structures and a given communicative function – especially, the referential or representative one –, but also translating numerous technical terms, whose meaning often escapes the translator's knowledge, whether this person is specialised or not, and whose translation is not always available in dictionaries and glossaries. On this basis, we propose that it can be highly useful to create ontologies for translators, as these provide conceptual and terminological information about a certain field of specialisation, by specifying its properties and connecting some concepts to others, so that they allow a better and deeper understanding of the subject requiring translation.

In this respect, an ontology, according to the Diccionario de la Real Academia Española (Dictionary of the Royal Spanish Academy), is the "part of the metaphysics that deals with the being in general and with its transcendental properties".[2] In the field of Knowledge Engineering, a discipline that has studied ontologies intensively,[3]

1 The research reported in this paper has been carried out in the framework of an R&D Project for Excellence *La contratación turística electrónica multilingüe como mediación intercultural: aspectos legales, traductológicos y terminológicos* [Multilingual tourism e-contracts: legal, translational and terminological aspects]. Funding source: Andalusian Ministry of Education, Science and Technology. Ref. no. HUM-892 (2006-2009). Project leader: G. Corpas Pastor.

2 The original definition is as follows: "Parte de la metafísica que trata del ser en general y de sus propiedades trascendentales".

3 Ontologies are a medium to create knowledge-based systems, so that knowledge can be codified and processed by computer.

Gruber's definition (1993: 199) is usually cited: "an explicit specification of a conceptualization". Weigand (1997: 138), for his part, offers a more specific definition: "an ontology is a database describing the concepts in the world or some domain, some of their properties, and how the concepts relate to each other".

In the field of definitions, it is also worth citing Studer *et al.* (1998: 25), who have expanded the well-known definition by Gruber and consider that an ontology is a formal, explicit specification of a shared conceptualisation. In order to explain their definition, the authors specify the meaning of each one of its elements:

> A 'conceptualisation' refers to an abstract model of some phenomenon in the world by having identified the relevant concepts of that phenomenon. 'Explicit' means that the type of concepts used, and the constraints on their use are explicitly defined. For example, in medical domains, the concepts are diseases and symptoms, the relations between them are causal and a constraint is that a disease cannot cause itself. 'Formal' refers to the fact that the ontology should be machine readable, which excludes natural language. 'Shared' reflects the notion that an ontology captures consensual knowledge, that is, it is not private to some individual, but accepted by a group.

Any ontology consists of a series of components. Noy/McGuinness (2001) cite the following:

- Classes, sometimes called concepts, which make up a domain of discourse.
- Properties of each concept, also called slots, which describe various features and attributes of the concept.
- Restrictions on slots, also called facets, which specify the values and the type of values that the properties can take.
- Instances, which are individual elements of the ontology.

Corcho *et al.* (2005: 144-145), for their part, propose a wider composition and distinguish between concepts, relations, instances, constants, attributes, axioms, and rules:

- Concepts represent ideas about the physical or abstract objects that constitute a domain. As the authors point out, concepts are usually organised in taxonomies through which inheritance mechanisms can be applied. In this way, a class can be divided into subclasses that represent more specific concepts: for instance, in the domain we are dealing with, an ontology about diabetes can contain the class "diabetes mellitus", and this can further contain the subclass "type 1 diabetes".

- Relations represent a type of association between concepts of the domain. The majority of relations link two concepts, so they are called binary relations.
- Instances represent individuals or specific elements of an ontology. So, for instance, the concept "insulin" can be instantiated as "insulin aspart".
- Constants are numerical values that do not change during much time. For example, the fasting blood glucose level, from which diabetes is diagnosed, is 126 mg/dL.
- Attributes describe properties of concepts and of instances. The authors distinguish two types of attributes: class attributes and instance attributes. The former ones describe concepts and take their values in the concept where they are defined. The latter ones describe instances and take their values in them.
- Formal axioms are logical expressions that are always true and are normally used to specify constraints in the ontology.
- Rules are generally used to infer knowledge in the ontology, such as attribute values.

Our study will be based on the composition proposed by Corcho *et al.* (2005). On the other hand, ontologies can be classified according to several criteria. We will mention here Gómez-Pérez *et al.* (2003), Guarino (1998) and Steve *et al.* (1998), since they have offered clear and widespread definitions in the field we are dealing with.

According to Gómez-Pérez *et al.* (2003: 8), ontologies can be classified into two groups, *lightweight* and *heavyweight* ontologies: the former ones include concepts, concept taxonomies, relationships between concepts, and properties that describe concepts; on the other side, heavyweight ontologies contain, apart from the previous elements, axioms and constraints, which clarify the intended meaning of the terms gathered on the ontology.

For his part, Guarino (1998: 7-8) distinguishes four large ontology groups, according to their level of generality:

- Top-level ontologies describe very general concepts like space, time, matter, object, event, action, etc., which are independent of a particular problem or domain.
- Domain ontologies describe the vocabulary related to a generic domain (like medicine, or automobiles).
- Task ontologies describe a generic task or activity (like diagnosing or selling).
- Application ontologies describe concepts depending both on a particular domain and task, which are often specialisations of both the related ontologies.

Steve *et al.* (1998: 1-2) propose a similar classification to the one by Guarino; like this author, they include domain ontologies, whereas they divide top-level ontologies into generic and representation ones:

- Domain ontologies concern specialised knowledge in a domain or subdomain: medicine, engineering, cardiology, law, etc.
- Generic ontologies concern general, foundational aspects of knowledge: processes, part/whole structure, kinds of objects, or quantities.
- Representation ontologies specify the conceptualisations that underlie knowledge representation formalisms, so they are also called metaontologies (*meta-level* or *top-level ontologies*).

The ontology created in the scope of this study will be of the domain type, as it is defined by Guarino (1998) and Steve *et al.* (1998), and of the lightweight type.

Finally, within the introductory aspects, we would also like to take into consideration the relationship between ontologies and terminology, since, as Vargas (2007: 50-51) states, they have certain points in common, given that a substantial part of the terminologist's work needs a conceptual approach. In this way, it is necessary to understand the area being explored in order to be able to structure it, classify it and define it. In this respect, the author states that a conceptual structure must allow for the following:

1) To circumscribe the research, since the fields and subfields the area is divided into will be considered.
2) To anticipate both the set of relevant terms of the domain being explored, and those that belong to related areas.
3) To determine the relevance of the terms in the domain; by dividing them into groups related to a certain subfield, it is possible to establish which ones are really interesting, based on the goals and the prototype user of the terminological application to be created (database, dictionary, vocabulary, etc.).
4) To construct a more controlled and coherent terminology, given that structuring makes it possible to treat an area of specialty in a more systematic way.
5) To define terminological units in a logical and systematic way.

Vargas (2007: 48-49) also mentions some projects that have taken into account the relationship between ontologies and terminology: for example, the COGNITERM Project about terminological knowledge bases, developed at the University of Ottawa and described in Meyer *et al.* (1992), can be considered as the pioneer in addressing the interaction between terminology and knowledge representation through ontologies. In the Spanish context, for its part, the ONCOTERM Project was among the first ones to address the aforementioned approach, as it combined techniques for information analysis and representation based on ontological semantics (see Nirenburg/Raskin 2001) with corpus linguistics (Faber 2002), in order to structure the oncology subdomain and create a multilingual information system.

In our study we will continue and expand this line of research that combines terminology and ontologies, and we will present a new way to build onto terminological resources, in this case aimed at scientific-technical translators. Thus, we consider that an ontology can provide the translator with the following information: a) the translation of a term; b) how that term is related to others from the same domain; and c) what properties and characteristics it has. Works such as the Dictionnaire Analytique de la Distribution by Dancette and Rhétoré (2000) have applied this translation-oriented approach in a similar way.

2. Construction of an ontology about the diabetes domain

As we pointed out in the previous section, we start from the hypothesis that ontologies can become a valuable medium of conceptual and terminological documentation for the translator, as they are capable of providing non-ambiguous information about the concepts of a domain in several languages, thus facilitating information transfer worldwide.

The first step in the construction of an ontology involves determining its domain and its scope. In this respect, Noy/McGuinness (2001) propose some questions to answer prior to its creation:

- What is the domain that the ontology will cover?
- For what are we going to use the ontology?
- For what types of questions should the information in the ontology provide answers?
- Who will use and maintain the ontology?

In this regard, the domain will be diabetes, covering at first general concepts from informative works aimed at the general public. The ontology will be used for terminological, translational and documentation purposes and will thus respond to questions that translators and terminologists could pose regarding informative texts from the domain of diabetes. It will contain linguistic realisations, or lexicalisations, in English, German and Spanish. Its maintenance will be initially undertaken by the researcher responsible for this study, but this task could be later on carried out by translators and terminologists, with an appropriate computer application.

This first step is part of a methodology for constructing ontologies proposed by Noy and McGuinness (2001), which is at the same time linked to the tool Protégé-2000. According to this methodology, after determining the domain, we must consider reusing existing ontologies; then, we must enumerate the important terms in the ontology, define the classes and the class hierarchy, define the properties of the classes and their facets, and define instances. Gómez-Pérez *et al.* (2003: 113-142) mention other relevant methods (Cyc, Uschold and King, Grüninger and Fox, KACTUS) and describe a methodology, METHONTOLOGY, created within the Ontological Engineering Group of the Universidad Politécnica de Madrid (UPM).

For our part, in order to illustrate and demonstrate the aforementioned hypothesis, we have resorted to the Termontography approach (Temmerman/Kerremans 2003) and to the methodology for ontology creation METHONTOLOGY (Gómez-Pérez/Fernández-López/Corcho 2003).

Termontography (Temmerman/Kerremans 2003: 4) is an approach – developed by the research group of the Centrum voor Vaktaal en Communicatie (CVC) in Brussels – which combines ontologies with the theories and methods for multilingual terminological analysis of the sociocognitive approach (Temmerman 2000). This theory stands out especially because of its capacity to represent the lexical gaps of certain languages, as well as the concepts that are particular to one language only. With this aim, "the termontography method works middle-out" (Temmerman/Kerremans 2003: 4), that is, it combines top-down and bottom-up approaches. Firstly, in close collaboration with experts of the domain in question, an initial framework of categories and inter-categorical relationships is developed top-down. Next, as the authors explain, this categorisation framework serves as a template for the manual and semi-automatic extraction of knowledge from a corpus. Afterwards, as the textual information is extracted from the corpus and this is compared to the categorical framework, the categories, even the culture-specific ones, are gradually expanded, as well as the semantic relationships among them, so a bottom-up analysis takes place. In this way, we can observe the cultural, conceptual and linguistic differences that may exist within the same domain. The results of this analysis are finally reflected in a termontological database.

We have chosen this methodology because, together with METHONTOLOGY, it suits our goals completely: it starts from premises compatible with ours and presents some techniques that permit us to create an ontology for terminology and translation. Thus, the first step of our work has been the development of an initial framework of categories and inter-categorical relations. This is an initial structure, or basic ontology, which contains the main concepts of the domain that we are studying, diabetes. It is at this point when we have introduced the other methodology mentioned above: METHONTOLOGY.

The METHONTOLOGY method, described in Gómez-Pérez/Fernández-López/Corcho (2003), comes from the field of Knowledge Engineering – at the same time branch of Artificial Intelligence – and involves carrying out a series of steps, each of which specifies a different element of the ontology: building a glossary of terms, building concept taxonomies, building a concept dictionary, and defining the following elements in detail: ad hoc binary relations, instance attributes, class attributes, constants, formal axioms, rules, and instances. Given below are the steps that we have followed to develop our initial framework.

1) Building a glossary of terms

In the first place we have made a glossary of terms about diabetes. To this end, we have created a small corpus made up of informative texts and encyclopaedic entries, in Spanish, English and German.[4] In this way, we extracted the most relevant terms of the domain from the Spanish texts (concepts, attributes, relations, etc.), and we specified data such as their description and their typology. For this task, the Diabetes Dictionary of the National Diabetes Information Clearinghouse (United States), which has a Spanish version, was a great help.[5] Next, the texts in English and German were analysed, in search of the equivalents of the terms found in Spanish; the equivalents were then introduced in the column corresponding to "Synonyms". Afterwards, we checked whether our terms in English and German had the same meanings as the Spanish ones, so we searched for their definitions in dictionaries. We show here an extract of the glossary of terms (Table 1)[6]:

4 Some of the most relevant resources that we have used to create the corpus are the Web sites of MedlinePlus (<http://www.nlm.nih.gov/medlineplus/diabetes.html>), with texts in English and Spanish; Sociedad Española de Diabetes (Spanish Diabetes Society), with texts in English, German and Spanish (<http://www.sediabetes.org/web/contenidogeneral.asp?id=1094>); the American Diabetes Association (<http://www.diabetes.org>), with texts in English and Spanish; and the portal site DiabetesDE (<http//www.diabetesde.org>), with information in German.

5 Available at <http://diabetes.niddk.nih.gov/dm/pubs/dictionary/>.

6 This is an English version of our glossary.

Name	Synonyms	Description	Type
hyperglycemia	(ES) hiperglucemia (DE) Hyperglykämie	Excessive blood glucose. Fasting hyperglycemia is blood glucose above a desirable level after a person has fasted for at least 8 hours. Postprandial hyperglycemia is blood glucose above a desirable level 1 to 2 hours after a person has eaten.	Class attribute
insulin	(ES) insulina (DE) Insulin	A hormone that helps the body use glucose for energy. The beta cells of the pancreas make insulin. When the body cannot make enough insulin, insulin is taken by injection or through use of an insulin pump.	concept
glucagon	(ES) glucagón (DE) Glukagon	A hormone produced by the alpha cells in the pancreas. It raises blood glucose. An injectable form of glucagon, available by prescription, may be used to treat severe hypoglycemia.	concept

Table 1. Excerpt of the glossary of terms.

2) Building concept taxonomies

In the second place, we selected the concepts from the previous list and defined concept hierarchies. To this end, not only the Dictionary we mentioned above was a great help, but also the concept hierarchy available at the lexical database WordNet.[7] METHONTOLOGY distinguishes four types of taxonomies: Subclass-Of, Disjoint-Decomposition, Exhaustive-Decomposition, and Partition.

Graph 1 shows an example of partition, which consists of a set of subclasses that does not share common instances but that covers all the possible subclasses of the main concept where they come from (in this case, "diabetes mellitus"). Therefore, there are not instances of "diabetes mellitus" that are not instances of one of the subclasses in the partition.

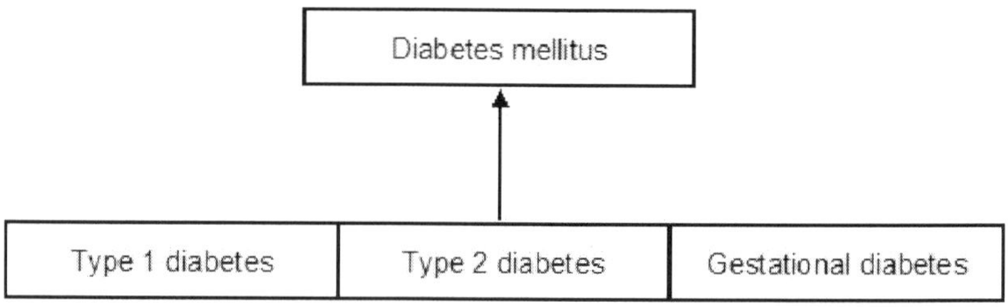

Graph 1. Example of partition.

3) Building ad hoc binary relation diagrams

The goal of these diagrams, according to Gómez-Pérez *et al.* (2003), is to establish ad hoc relationships between concepts of the same (or different) concept taxonomy. These are exclusive relationships in a certain domain. Graph 2 shows the relation "produces" and its inverse, "is produced by".

7 Available at <http://wordnet.princeton.edu>.

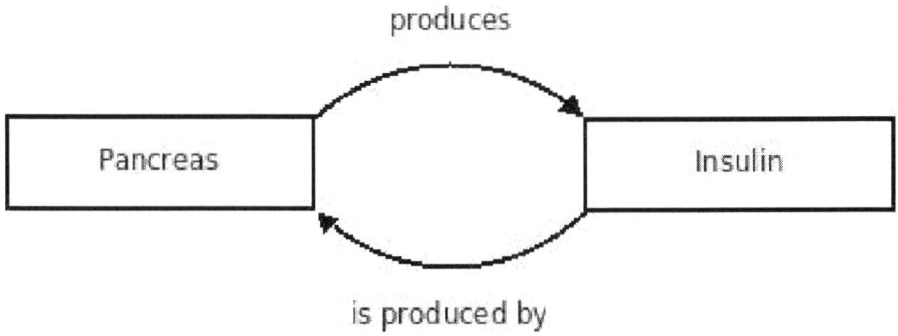

Graph 2. Example of ad hoc relation.

4) Building the concept dictionary

It contains all the concepts of the ontology, as well as the properties – both on a concept level and on an instance level – and the relations that describe them. It is possible that some concepts do not have class or instance attributes. Table 2 shows an excerpt of our concept dictionary.

Concept name	Class attributes	Instance attributes	Relations
diabetes	polyuria	--	partition: diabetes mellitus, diabetes insipidus
diabetes mellitus	hyperglycemia	--	partition: type 1 diabetes, type 2 diabetes, gestational diabetes
insulin	protein structure	area of injection retardant	is produced by (pancreas) Subclass-Of: rapid-acting insulin, regular or short-acting insulin, intermediate-acting insulin, long-acting insulin, premixed insulin

Table 2. Excerpt of the concept dictionary.

5) Defining ad hoc binary relations in detail

This step involves taking all the ad hoc binary relations included in the concept dictionary, detailing the name of the relation, the names of the source and target concepts, the cardinality – in this case, the number of elements that can act as a source concept –, and the possible inverse relations (see Table 3).

Relation name	Source concept	Source cardinality (max.)	Target concept	Inverse relation
produces	pancreas	1	insulin	is produced by
raises in blood	glucagon	1	glucose	is raised in blood
indicates	hemoglobin A1C	1	average blood glucose level (during previous months)	is indicated by

Table 3. Excerpt of the ad hoc binary relation table.

6) Defining instance attributes in detail

In this step we describe all the instance attributes included in the concept dictionary (see Table 4). The value type refers to whether the value is a string, an integer, boolean-type, etc. The range of values stands for the minimum and maximum values that the numerical attributes can take. For its part, the cardinality stands for the minimum and maximum number of values that the attribute can take.

Instance attribute name	Concept name	Value type	Range of values	Cardinality
area of injection	insulin	String	--	(1, N)
retardant	insulin	Yes/No	--	(1, 1)

Table 4. Excerpt of the instance attribute table.

7) Defining the class attributes in detail

In this phase we describe the class attributes reflected in the concept dictionary (see Table 5).

Class attributes	Concept name	Value type	Cardinality	Values
hyperglycemia	diabetes mellitus	Yes/No	(1, 1)	Yes
polyuria	diabetes	Yes/No	(1, 1)	Yes
protein structure	insulin	Yes/No	(1, 1)	Yes

Table 5. Excerpt of the class attribute table.

8) Defining the constants in detail

In this phase we define the constants which appear in the glossary of terms (see Table 6).

Name	Value type	Value	Measurement unit
Fasting blood glucose level, from which diabetes is diagnosed	integer	126	mg/dL

Table 6. Excerpt of the constant table.

9) Defining instances

Finally, we have defined some relevant instances of the diabetes domain. Table 7 includes an excerpt with two examples:

Instance name	Concept name	Attribute	Values
insulin aspart	rapid-acting insulin	area of injection	subcutaneous area
insulin lispro	rapid-acting insulin	area of injection	subcutaneous area intramuscular injection

Table 7. Examples of instances.

Although the starting language for this study – both for the conceptual formalisation of the domain and for the first linguistic realisations of those concepts – has been Spanish, this conceptualisation is likewise applicable to other countries and cultural areas, since we are dealing with a scientific field. On this basis, it has proven straightforward to establish correspondences between the Spanish lexicalisations and the English and German ones. In this respect, the glossary of terms that we initially created has been highly useful, since it has allowed us to quickly find the terms equivalent to those that we had already selected in Spanish. On the other hand, we have not included formal axioms and rules yet in this version, so, at this stage of development, we could consider our ontology to be lightweight, as it contains concepts, taxonomies, relations and properties.

According to the proposed method, once we have obtained the categorisation framework by means of the application of METHONTOLOGY, we take up again the use of Termontography. Our work is still under development, although we have started to carry out the following tasks:

- In the first place, we are extending the corpus with more informative texts and we have begun to introduce specialised texts.
- In the second place, the categorisation framework is being applied as a template for the manual extraction of concepts from the corpus. Thus, the existing concepts in the framework are helping us to find other related concepts, as well as new relations and attributes. Moreover, we are exploring the possibilities to carry out a semi-automatic extraction, for which some techniques already exist.

- In the third place, rules and formal axioms will be defined.
- Lastly, in order to process the information by computer and use it for our purposes, we will proceed to introduce all the concepts, relations, properties and axioms in the WebODE online platform, developed within the Ontological Engineering Group of the UPM.

Significantly, the methodology Termontography, given its capacity to represent lexical gaps and exclusive concepts, will allow us to also determine whether there are conceptual differences within the field of diabetes in the several countries whose languages are involved in the study.

3. Conclusions

In this article, we have demonstrated that it is possible to build an ontology, in a relatively simple way, for terminological and translational purposes. In this way, we have observed that the data included in the ontology can be highly useful for the translator when it comes to understanding the concepts of a certain domain and finding the most appropriate translation for the terms of a source text.

Moreover, we have observed that the methodologies METHONTOLOGY and Termontography can be perfectly applied for these purposes, since their combination permits to create solid ontoterminological resources, based on the techniques used by Knowledge Engineering and on the ontology-based terminological methods that Temmerman and her research group have been applying for a number of years, with remarkable diffusion.

Although this research is still under development, we consider that it will give rise to an ontoterminological resource of high quality for translators, because, given a term, it will provide information about the corresponding concept, as well as about its characteristics, properties, and relationships to other concepts; all of the above will be in turn bound to their corresponding lexicalisations, in Spanish, English and German.

Finally, we must highlight that the methodology described in this study is applicable to other specialty domains, so it is not limited only to the scientific-technical field.

References

Corcho, Óscar / Fernández-López, Mariano / Gómez-Pérez, Asunción / López-Cima, Ángel 2005. Building Legal Ontologies with METHONTOLOGY and WebODE. In Benjamins, Richard / Casanovas, Pompeu / Breuker, Joost /Gangemini, Aldo (eds) *Law and the Semantic Web*. Heidelberg: Springer Verlag, 142-157.

Dancette, Jeanne / Rhétoré, Christophe 2000. *Dictionnaire Analytique de la Distribution*. Montreal: Les Presses de l'Université de Montréal.

Faber, Pamela 2002. Investigar en terminología. In Faber, Pamela / Jiménez, Catalina (eds) *Investigar en Terminología*. Granada: Comares.

Gómez-Pérez, Asunción / Fernández-López, Mariano / Corcho, Óscar 2003. *Ontological Engineering*. Springer Verlag.

Gruber, Tom 1993. A Translation Approach to Portable Ontology Specifications. *Knowledge Acquisition*. 5/2, 199-220.

Guarino, Nicola 1998. Formal Ontology in Information Systems. In Guarino, Nicola (ed.) *Formal Ontology in Information Systems. Proceedings of FOIS'98, Trento, Italy, 6-8 June 1998*. Amsterdam: IOS Press, 3-15.

Meyer, Ingrid / Bowker, Lynne / Eck, Karen 1992. COGNITERM: An Experiment in Building a Terminological Knowledge Base. In Tommola, Hannu *et al.* (eds) *EURALEX '92 Proceedings*. Tampere: University of Tampere.

Nirenburg, Sergei / Raskin, Victor 2001. *Ontological Semantics* [electronic book]. <http://crl.nmsu.edu/Staff.pages/Technical/sergei/book/index-book.html>

Noy, Natalya / McGuinness, Deborah 2001. Ontology Development 101: A Guide to Creating Your First Ontology. Stanford Knowledge Systems Laboratory Technical Report KSL-01-05 and Stanford Medical Informatics Technical Report SMI-2001-0880. <http://ksl.stanford.edu/people/dlm/papers/ontology-tutorial-noy-mcguinness-abstract.html>

Real Academia Española 2001. *Diccionario de la Lengua Española* (DRAE). Madrid: Espasa.

Steve, Geri / Gangemi, Aldo / Pisanelli, Domenico 1998. *Integrating Medical Terminologies with ONIONS Methodology*. Technical Report ITBM-RIM-3/1997. Laboratory for Applied Ontology, Institute of Cognitive Sciences and Technology, Italy.

Studer, Rudi / Benjamins, Richard / Fensel, Dieter 1998. Knowledge Engineering: Principles and Methods. *Data and Knowledge Engineering*. 25, 161-197.

Temmerman, Rita 2000. *Towards New Ways of Terminology Description. The Sociocognitive Approach*. Amsterdam/Philadelphia: John Benjamins.

Temmerman, Rita / Kerremans, Koen 2003. Termontography: Ontology Building and the Sociocognitive Approach to Terminology Description. In Hajicová, Eva / Kotešovcová, Anna / Mírovský, Jiří (eds) *Proceedings of CIL17*. Prag, Czech Republic: Matfyzpress, MFF UK.

Vargas, Chelo 2007. El léxico especializado y las ontologías. In Alcaraz Varó, Enrique / Mateo Martínez, José / Yus Ramos, Francisco (eds) *Las lenguas profesionales y académicas*. Barcelona: Ariel.

Weigand, Hans 1997. Multilingual Ontology-Based Lexicon for News Filtering – The TREVI Project. In Mahesh, Kavi (ed.) *Ontologies and Multilingual NLP. Proceedings from the 15th International Joint Conference on Artificial Intelligence (IJCAI), August 23-29, 1997*. Nagoya, Japan, 138-159.

Notes on Contributors

María Rosario Bautista Zambrana is a lecturer in German at the Department of English, French and German Philology at the University of Málaga. She obtained an MA in Translation and Interpreting (English and German) at the University of Málaga in 2004. She is a member of the research group *Lexicography and Translation (HUM-106)* in the same university, and has taken part in a number of research and education innovation projects. She is currently writing her PhD thesis on package travel and ontologies.

José Ramón Biau Gil is a freelance translator and a lecturer in English and Translation at the Universitat Rovira i Virgili (Tarragona, Spain). He holds a degree in English Language and Literature, and an MA in Technical Translation from the same university. He has published several works on translation technologies and related training.

Gonzalo Camina is a language teacher at the School of Arts, Literatures and Cultures at the University of Ulster, and a research fellow at the University of A Coruña. His main research areas include Early Modern English and corpus linguistics. His PhD dissertation (forthcoming) explores morphological issues of word formation in scientific English in the eighteenth century. He is a member of the *MuStE* research group, and has contributed to the compilation of the *Coruña Corpus*.

Pilar Cid Leal is Senior Lecturer in the Department of Translation and Interpreting at the Universitat Autònoma de Barcelona (UAB), where she conducts research on Information Science applied to Translation. She is the author of several papers and is a member of the editorial board of the journal *Tradumàtica*.

Tomás Conde has recently completed his PhD on Translation and Interpreting at the University of Granada, where he has been a member of the *PETRA* research group since 2002. His areas of interest include

She is currently working on her PhD thesis on the language of architecture and construction engineering textbooks. Her research activity mainly focuses on specialised writing, genre analysis and discourse analysis.

Mariana Orozco is a professional translator and Senior Lecturer in Translation at the School of Translation at the Universitat Autònoma de Barcelona (UAB). She holds a PhD in Translation Theory and has developed a measuring instrument to test the effects of different teaching methodologies on translation students. She has been involved in several funded research projects and has published more than twenty articles and book chapters on translation.

Françoise Salager-Meyer is Senior Lecturer at the Universidad de los Andes (Mérida, Venezuela), where she teaches ESP and Russian. Educated at the University of Lyon, France, and the University of Texas at Austin, she is the author of numerous publications on written medical discourse. Both in 1994 and in 2004, she was awarded the Horowitz Prize for her works on the pragmatics of written scholarly communication. She created, and is now coordinating, the *Multilingual and Multidisciplinary Research Group on Scientific Discourse Analysis* (Universidad de Los Andes).

Estefanía Sánchez Barreiro obtained an MA in English Language from the University of A Coruña in 2004. She has been a research assistant for the *MuStE* group since 2004. Her main research interests concern word-formation processes in the History of English and the diachronic study of extenders in scientific English. She is currently working on her PhD thesis on the use of extenders in English scientific writing.

Pilar Sánchez-Gijón is a lecturer in the Translation Department at the Universitat Autònoma de Barcelona (UAB), where she teaches Terminology and Computer Sciences applied to Translation Studies. She graduated in Translation and holds a Master's Degree on Translation Theory. Her PhD dealt with the use of Corpus Linguistics in the translation of specialised texts. Her research lines are terminology management and didactics, translation technologies and didactics, and corpus linguistics in translation. She has written several papers and a book on these topics.

Francisca Suau-Jiménez is Senior Lecturer in English Philology at the Universitat de València (Spain), and Director of the *Institut Interuniversitari de Llengües Modernes Aplicades (IULMA)* in the same university. Her research interests are discourse analysis in professional and scientific genres, and specialised translation. Her current projects involve modern languages applied to real industrial and academic demands, and corpus linguistics.

Olga Torres-Hostench obtained a PhD in Translation Theory from the Universitat Autònoma de Barcelona (UAB). She has been working on multimedia localisation since 1995. At present, she teaches translation at the UAB, and also web localisation in different postgraduate courses. She is a member of *Tradumàtica*, a UAB research group on new technologies applied to translation.

Linguistic Insights

Studies in Language and Communication

This series aims to promote specialist language studies in the fields of linguistic theory and applied linguistics, by publishing volumes that focus on specific aspects of language use in one or several languages and provide valuable insights into language and communication research. A cross-disciplinary approach is favoured and most European languages are accepted.

The series includes two types of books:

- **Monographs** – featuring in-depth studies on special aspects of language theory, language analysis or language teaching.
- **Collected papers** – assembling papers from workshops, conferences or symposia.

Each volume of the series is subjected to a double peer-reviewing process.

Vol. 1 Maurizio Gotti & Marina Dossena (eds)
Modality in Specialized Texts. Selected Papers of the 1st CERLIS Conference.
421 pages. 2001. ISBN 3-906767-10-8. US-ISBN 0-8204-5340-4

Vol. 2 Giuseppina Cortese & Philip Riley (eds)
Domain-specific English. Textual Practices across Communities and Classrooms.
420 pages. 2002. ISBN 3-906768-98-8. US-ISBN 0-8204-5884-8

Vol. 3 Maurizio Gotti, Dorothee Heller & Marina Dossena (eds)
Conflict and Negotiation in Specialized Texts. Selected Papers of the 2nd CERLIS Conference.
470 pages. 2002. ISBN 3-906769-12-7. US-ISBN 0-8204-5887-2

Editorial address:

Prof. Maurizio Gotti Università di Bergamo, Facoltà di Lingue e Letterature Straniere,
Via Salvecchio 19, 24129 Bergamo, Italy
Fax: 0039 035 2052789, E-Mail: m.gotti@unibg.it

Vol. 4 Maurizio Gotti, Marina Dossena, Richard Dury, Roberta Facchinetti & Maria Lima
Variation in Central Modals. A Repertoire of Forms and Types of Usage in Middle English and Early Modern English.
364 pages. 2002. ISBN 3-906769-84-4. US-ISBN 0-8204-5898-8

Vol. 5 Stefania Nuccorini (ed.)
Phrases and Phraseology. Data and Descriptions.
187 pages. 2002. ISBN 3-906770-08-7. US-ISBN 0-8204-5933-X

Vol. 6 Vijay Bhatia, Christopher N. Candlin & Maurizio Gotti (eds)
Legal Discourse in Multilingual and Multicultural Contexts. Arbitration Texts in Europe.
385 pages. 2003. ISBN 3-906770-85-0. US-ISBN 0-8204-6254-3

Vol. 7 Marina Dossena & Charles Jones (eds)
Insights into Late Modern English. 2nd edition.
378 pages. 2003, 2007.
ISBN 978-3-03911-257-9. US-ISBN 978-0-8204-8927-8

Vol. 8 Maurizio Gotti
Specialized Discourse. Linguistic Features and Changing Conventions.
351 pages. 2003, 2005.
ISBN 3-03910-606-6. US-ISBN 0-8204-7000-7

Vol. 9 Alan Partington, John Morley & Louann Haarman (eds)
Corpora and Discourse.
420 pages. 2004. ISBN 3-03910-026-2. US-ISBN 0-8204-6262-4

Vol. 10 Martina Möllering
The Acquisition of German Modal Particles. A Corpus-Based Approach.
290 pages. 2004. ISBN 3-03910-043-2. US-ISBN 0-8204-6273-X

Vol. 11 David Hart (ed.)
English Modality in Context. Diachronic Perspectives.
261 pages. 2003. ISBN 3-03910-046-7. US-ISBN 0-8204-6852-5

Vol. 12 Wendy Swanson
Modes of Co-reference as an Indicator of Genre.
430 pages. 2003. ISBN 3-03910-052-1. US-ISBN 0-8204-6855-X

Vol. 13 Gina Poncini
Discursive Strategies in Multicultural Business Meetings.
2nd edition. 338 pages. 2004, 2007.
ISBN 978-3-03911-296-8. US-ISBN 978-0-8204-8937-7

Vol. 14 Christopher N. Candlin & Maurizio Gotti (eds)
Intercultural Aspects of Specialized Communication.
2nd edition. 369 pages. 2004, 2007.
ISBN 978-3-03911-258-6. US-ISBN 978-0-8204-8926-1

Vol. 15 Gabriella Del Lungo Camiciotti & Elena Tognini Bonelli (eds)
Academic Discourse. New Insights into Evaluation.
234 pages. 2004. ISBN 3-03910-353-9. US-ISBN 0-8204-7016-3

Vol. 16 Marina Dossena & Roger Lass (eds)
Methods and Data in English Historical Dialectology.
405 pages. 2004. ISBN 3-03910-362-8. US-ISBN 0-8204-7018-X

Vol. 17 Judy Noguchi
The Science Review Article. An Opportune Genre in the Construction of Science.
274 pages. 2006. ISBN 3-03910-426-8. US-ISBN 0-8204-7034-1

Vol. 18 Giuseppina Cortese & Anna Duszak (eds)
Identity, Community, Discourse. English in Intercultural Settings.
495 pages. 2005. ISBN 3-03910-632-5. US-ISBN 0-8204-7163-1

Vol. 19 Anna Trosborg & Poul Erik Flyvholm Jørgensen (eds)
Business Discourse. Texts and Contexts.
250 pages. 2005. ISBN 3-03910-606-6. US-ISBN 0-8204-7000-7

Vol. 20 Christopher Williams
Tradition and Change in Legal English. Verbal Constructions in Prescriptive Texts.
2nd revised edition. 216 pages. 2005, 2007. ISBN 978-3-03911-444-3.

Vol. 21 Katarzyna Dziubalska-Kołaczyk & Joanna Przedlacka (eds)
English Pronunciation Models: A Changing Scene.
2nd edition. 476 pages. 2005, 2008. ISBN 978-3-03911-682-9.

Vol. 22 Christián Abello-Contesse, Rubén Chacón-Beltrán, M. Dolores López-Jiménez & M. Mar Torreblanca-López (eds)
Age in L2 Acquisition and Teaching.
214 pages. 2006. ISBN 3-03910-668-6. US-ISBN 0-8204-7174-7

Vol. 23 Vijay K. Bhatia, Maurizio Gotti, Jan Engberg & Dorothee Heller (eds)
Vagueness in Normative Texts.
474 pages. 2005. ISBN 3-03910-653-8. US-ISBN 0-8204-7169-0

Vol. 24 Paul Gillaerts & Maurizio Gotti (eds)
Genre Variation in Business Letters. 2nd printing.
407 pages. 2008. ISBN 978-3-03911-681-2.

Vol. 25 Ana María Hornero, María José Luzón & Silvia Murillo (eds)
Corpus Linguistics. Applications for the Study of English.
2nd printing. 526 pages. 2006, 2008. ISBN 978-3-03911-726-0

Vol. 26 J. Lachlan Mackenzie & María de los Ángeles Gómez-González (eds)
Studies in Functional Discourse Grammar.
259 pages. 2005. ISBN 3-03910-696-1 / US-ISBN 0-8204-7558-0

Vol. 27 Debbie G. E. Ho
Classroom Talk. Exploring the Sociocultural Structure of Formal ESL Learning.
2nd edition. 254 pages. 2006, 2007. ISBN 978-3-03911-434-4

Vol. 28 Javier Pérez-Guerra, Dolores González-Álvarez, Jorge L. Bueno-Alonso & Esperanza Rama-Martínez (eds)
'Of Varying Language and Opposing Creed': New Insights into Late Modern English.
455 pages. 2007. ISBN 978-3-03910-788-9

Vol. 29 Francesca Bargiela-Chiappini & Maurizio Gotti (eds)
Asian Business Discourse(s).
350 pages. 2005. ISBN 3-03910-804-2 / US-ISBN 0-8204-7574-2

Vol. 30 Nicholas Brownlees (ed.)
News Discourse in Early Modern Britain. Selected Papers of CHINED 2004.
300 pages. 2006. ISBN 3-03910-805-0 / US-ISBN 0-8204-8025-8

Vol. 31 Roberta Facchinetti & Matti Rissanen (eds)
Corpus-based Studies of Diachronic English.
300 pages. 2006. ISBN 3-03910-851-4 / US-ISBN 0-8204-8040-1

Vol. 32 Marina Dossena & Susan M. Fitzmaurice (eds)
Business and Official Correspondence: Historical Investigations.
209 pages. 2006. ISBN 3-03910-880-8 / US-ISBN 0-8204-8352-4

Vol. 33 Giuliana Garzone & Srikant Sarangi (eds)
Discourse, Ideology and Specialized Communication.
494 pages. 2007. ISBN 978-3-03910-888-6

Vol. 34 Giuliana Garzone & Cornelia Ilie (eds)
The Use of English in Institutional and Business Settings.
An Intercultural Perspective.
372 pages. 2007. ISBN 978-3-03910-889-3

Vol. 35 Vijay K. Bhatia & Maurizio Gotti (eds)
Explorations in Specialized Genres.
316 pages. 2006. ISBN 3-03910-995-2 / US-ISBN 0-8204-8372-9

Vol. 36 Heribert Picht (ed.)
Modern Approaches to Terminological Theories and Applications.
432 pages. 2006. ISBN 3-03911-156-6 / US-ISBN 0-8204-8380-X

Vol. 37 Anne Wagner & Sophie Cacciaguidi-Fahy (eds)
Legal Language and the Search for Clarity / Le langage juridique et la quête de clarté.
Practice and Tools / Pratiques et instruments.
487 pages. 2006. ISBN 3-03911-169-8 / US-ISBN 0-8388-5

Vol. 38 Juan Carlos Palmer-Silveira, Miguel F. Ruiz-Garrido &
Inmaculada Fortanet-Gómez (eds)
Intercultural and International Business Communication:
Theory, Research and Teaching.
2nd edition. 343 pages. 2006, 2008. ISBN 978-3-03911-680-5

Vol. 39 Christiane Dalton-Puffer, Dieter Kastovsky, Nikolaus Ritt &
Herbert Schendl (eds)
Syntax, Style and Grammatical Norms: English from 1500–2000.
250 pages. 2006. ISBN 3-03911-181-7 / US-ISBN 0-8204-8394-X

Vol. 40 Marina Dossena & Irma Taavitsainen (eds)
Diachronic Perspectives on Domain-Specific English.
280 pages. 2006. ISBN 3-03910-176-0 / US-ISBN 0-8204-8391-5

Vol. 41 John Flowerdew & Maurizio Gotti (eds)
Studies in Specialized Discourse.
293 pages. 2006. ISBN 3-03911-178-7

Vol. 42 Ken Hyland & Marina Bondi (eds)
Academic Discourse Across Disciplines.
320 pages. 2006. ISBN 3-03911-183-3 / US-ISBN 0-8204-8396-6

Vol. 43 Paul Gillaerts & Philip Shaw (eds)
The Map and the Landscape: Norms and Practices in Genre.
256 pages. 2006. ISBN 3-03911-182-5 / US-ISBN 0-8204-8395-4

Vol. 44 Maurizio Gotti & Davide Giannoni (eds)
New Trends in Specialized Discourse Analysis.
301 pages. 2006. ISBN 3-03911-184-1 / US-ISBN 0-8204-8381-8

Vol. 45 Maurizio Gotti & Françoise Salager-Meyer (eds)
Advances in Medical Discourse Analysis: Oral and Written Contexts.
492 pages. 2006. ISBN 3-03911-185-X / US-ISBN 0-8204-8382-6

Vol. 46 Maurizio Gotti & Susan Šarcević (eds)
Insights into Specialized Translation.
396 pages. 2006. ISBN 3-03911-186-8 / US-ISBN 0-8204-8383-4

Vol. 47 Khurshid Ahmad & Margaret Rogers (eds)
Evidence-based LSP: Translation, Text and Terminology.
584 pages. 2007. ISBN 978-3-03911-187-9

Vol. 48 Hao Sun & Dániel Z. Kádár (eds)
It's the Dragon's Turn: Chinese Institutional Discourses.
262 pages. 2008. ISBN 978-3-03911-175-6

Vol. 49 Cristina Suárez-Gómez
Relativization in Early English (950-1250): the Position of Relative Clauses.
149 pages. 2006. ISBN 3-03911-203-1 / US-ISBN 0-8204-8904-2

Vol. 50 Maria Vittoria Calvi & Luisa Chierichetti (eds)
Nuevas tendencias en el discurso de especialidad.
319 pages. 2006. ISBN 978-3-03911-261-6

Vol. 51 Mari Carmen Campoy & María José Luzón (eds)
Spoken Corpora in Applied Linguistics.
274 pages. 2008. ISBN 978-3-03911-275-3

Vol. 52 Konrad Ehlich & Dorothee Heller (Hrsg.)
Die Wissenschaft und ihre Sprachen.
323 pages. 2006. ISBN 978-3-03911-272-2

Vol. 53 Jingyu Zhang
The Semantic Salience Hierarchy Model: The L2 Acquisition of Psych Predicates
273 pages. 2007. ISBN 978-3-03911-300-2

Vol. 54 Norman Fairclough, Giuseppina Cortese & Patrizia Ardizzone (eds)
Discourse and Contemporary Social Change.
555 pages. 2007. ISBN 978-3-03911-276-0

Vol. 55 Jan Engberg, Marianne Grove Ditlevsen, Peter Kastberg & Martin Stegu (eds)
New Directions in LSP Teaching.
331 pages. 2007. ISBN 978-3-03911-433-7

Vol. 83 Victoria Guillén-Nieto, Carmen Marimón-Llorca & Chelo Vargas-Sierra (eds)
Intercultural Business Communication and
Simulation and Gaming Methodology.
392 pages. 2009. ISBN 978-3-03911-688-1

Vol. 84 Maria Grazia Guido
English as a Lingua Franca in Cross-cultural Immigration Domains.
285 pages. 2008. ISBN 978-3-03911-689-8

Vol. 85 Erik Castello
Text Complexity and Reading Comprehension Tests.
352 pages. 2008. ISBN 978-3-03911-717-8

Vol. 86 Maria-Lluisa Gea-Valor, Isabel García-Izquierdo & Maria-José Esteve (eds)
Linguistic and Translation Studies in Scientific Communication.
317 pages. 2010. ISBN 978-3-0343-0069-8

Vol. 87 Carmen Navarro, Rosa Mª Rodríguez Abella, Francesca Dalle Pezze
& Renzo Miotti (eds)
La comunicación especializada.
355 pages. 2008. ISBN 978-3-03911-733-8

Vol. 88 Kiriko Sato
The Development from Case-Forms to Prepositional Constructions
in Old English Prose.
231 pages. 2009. ISBN 978-3-03911-763-5

Vol. 89 Dorothee Heller (Hrsg.)
Formulierungsmuster in deutscher und italienischer Fachkommunikation.
Intra- und interlinguale Perspektiven.
315 pages. 2008. ISBN 978-3-03911-778-9

Vol. 90 Henning Bergenholtz, Sandro Nielsen & Sven Tarp (eds)
Lexicography at a Crossroads. Dictionaries and Encyclopedias Today,
Lexicographical Tools Tomorrow.
372 pages. 2009. ISBN 978-3-03911-799-4

Vol. 91 Manouchehr Moshtagh Khorasani
The Development of Controversies: From the Early Modern Period
to Online Discussion Forums.
317 pages. 2009. ISBN 978-3-3911-711-6

Vol. 92 María Luisa Carrió-Pastor (ed.)
Content and Language Integrated Learning: Cultural Diversity.
178 pages. 2009. ISBN 978-3-3911-818-2

Vol. 93-94 Forthcoming.

Vol. 95 Facchinetti Roberta, Crystal David, Seidlhofer Barbara (eds)
From International to Local English – And Back Again.
268 pages. 2010. ISBN 978-3-0343-0011-7

Vol. 96 Cesare Gagliardi & Alan Maley (eds)
EIL, ELF, Global English: Teaching and Learning Issues
376 pages. 2010. ISBN 978-3-0343-0010-0

Vol. 97 Sylvie Hancil (ed.)
The Role of Prosody in Affective Speech.
403 pages. 2009. ISBN 978-3-03911-696-6

Vol. 98 Marina Dossena & Roger Lass (eds)
Studies in English and European Historical Dialectology.
257 pages. 2009. ISBN 978-3-0343-0024-7

Vol. 99 Christine Béal
Les interactions quotidiennes en français et en anglais.
De l'approche comparative à l'analyse des situations interculturelles.
424 pages. 2010. ISBN 978-3-0343-0027-8

Vol. 100 Maurizio Gotti (ed.)
Commonality and Individuality in Academic Discourse.
398 pages. 2009. ISBN 978-3-0343-0023-0

Vol. 101 Javier E. Díaz Vera & Rosario Caballero (eds)
Textual Healing: Studies in Medieval English Medical, Scientific and Technical Texts.
213 pages. 2009. ISBN 978-3-03911-822-9

Vol. 102 Nuria Edo Marzá
The Specialised Lexicographical Approach: A Step further in Dictionary-making.
316 pages. 2009. ISBN 978-3-0343-0043-8

Vol. 103 Carlos Prado-Alonso, Lidia Gómez-García, Iria Pastor-Gómez & David Tizón-Couto (eds)
New Trends and Methodologies in Applied English Language Research. Diachronic, Diatopic and Contrastive Studies.
348 pages. 2009. ISBN 978-3-0343-0046-9

Vol.104-108 Forthcoming.

Vol. 109 Ángeles Linde López & Rosalía Crespo Jiménez (eds)
Professional English in the European context: The EHEA challenge.
374 pages. 2010. ISBN 978-3-0343-0088-9

Vol. 110 Rosalía Rodríguez-Vázquez
The Rhythm of Speech, Verse and Vocal Music: A New Theory.
394 pages. 2010. ISBN 978-3-0343-0309-5

Vol. 111 Anastasios Tsangalidis & Roberta Facchinetti (eds)
Studies on English Modality. In Honour of Frank Palmer.
392 pages. 2009. ISBN 978-3-0343-0310-1

Vol. 112-113 Forthcoming.

Vol. 114 Merja Kytö, John Scahill & Harumi Tanabe (eds)
Language Change and Variation from Old English to Late Modern English. A Festschrift for Minoji Akimoto
422 pages. 2010. ISBN 978-3-0343-0372-9

Vol. 115 Giuliana Garzone & Paola Catenaccio (eds)
Identities across Media and Modes: Discursive Perspectives.
379 pages. 2009. ISBN 978-3-0343-0386-6

Vol. 116 Elena Landone
Los marcadores del discurso y cortesía verbal en español.
390 pages. 2010. ISBN 978-3-0343-0413-9

Vol. 117 Maurizio Gotti & Christopher Williams (eds)
Legal Discourse across Languages and Cultures.
339 pages. 2010. ISBN 978-3-0343-0425-2

Vol. 118 David Hirsh
Academic Vocabulary in Context.
217 pages. 2010. ISBN 978-3-0343-0426-9

Vol. 119-120 Forthcoming.

Vol. 121 Davide Simone Giannoni & Celina Frade (eds)
Researching Language and the Law. Textual Features and Translation Issues.
278 pages. 2010. ISBN 978-3-0343-0443-6

Vol. 122 Forthcoming.

Vol. 123 Vijay K. Bhatia, Christopher N. Candlin & Maurizio Gotti (eds)
The Discourses of Dispute Resolution.
290 pages. 2010. ISBN 978-3-0343-0476-4

Vol. 124 Davide Simone Giannoni
Mapping Academic Values in the Disciplines. A Corpus-Based Approach.
288 pages. 2010. ISBN 978-3-0343-0488-7

Vol. 125 Giuliana Garzone & James Archibald (eds)
Discourse, Identities and Roles in Specialized Communication.
419 pages. 2010. ISBN 978-3-0343-0494-8